W0043280

Pathochemical Markers in Major Psychoses

Edited by
H. Beckmann and P. Riederer

With Contributions by
M. Ackenheil, M. Albus, M. Arato, C. M. Banki
H. Beckmann, W. Berrettini, H. J. Bochnik, B. Bondy
S. M. Bonham Carter, J. Bruinvels, L. Demisch, E. Gabriel
T. Gasser, W. F. Gattaz, A. Gjerris, K. Jellinger
D. C. Jimerson, D. P. van Kammen, W. B. van Kammen
M. Linnoila, L. S. Mann, E. A. Mueller, F. Müller-Spahn
U. Münch, D. L. Murphy, D. Naber, P. T. Ninan, Z. Papp
L. Pepplinkhuizen, O. J. Rafaelsen, F. Reinhuber
G. P. Reynolds, P. Riederer, M. Sandler, M. Scheinin
L. J. Siever

With 28 Figures

Springer-Verlag
Berlin Heidelberg New York Tokyo 1985

Professor Dr. HELMUT BECKMANN, Zentralinstitut für Seelische Gesundheit Mannheim, Psychiatrische Klinik, Postfach 59 70, 6800 Mannheim

Professor Dr. PETER RIEDERER, L. Boltzmann-Institut für Klinische Neurobiologie, Abteilung Neurochemie, Lainz Hospital, Wolkersbergenstraße 1, A-1130 Wien

ISBN-13:978-3-642-69745-6   e-ISBN-13:978-3-642-69743-2
DOI:10.1007/978-3-642-69743-2

Library of Congress Cataloging in Publication Data Main entry under title: Pathochemical markers in major psychoses. Proceedings of a symposium held in Vienna, Austria in July 1983. Includes index. 1. Psychoses–Physiological aspects–Congresses. 2. Biochemical markers–Congresses. 3. Brain chemistry–Congresses. I. Beckmann, H. (Helmut), 1940–. II. Riederer, P., 1942–. [DNLM: 1. Psychotic Disorders–physiopathology–congresses. WM 200 P297 1983] RC512.P297 1984   616.89   84-14195

This work is subject to copyright. All rights are reserved, whether the whole or part of the material is concerned, specifically those of translation, reprinting, re-use of illustrations, broadcasting, reproduction by photocopying machine or similar means, and storage in data banks. Under § 54 of the German Copyright Law where copies are made for other than private use a fee is payable to 'Verwertungsgesellschaft Wort', Munich.

© Springer-Verlag Berlin Heidelberg 1985
Softcover reprint of the hardcover 1st edition 18985

The use of registered names, trademarks, etc. in the publication does not imply, even in the absence of a specific statement, that such names are exempt from the relevant protective laws and regulations and therefore free for general use.

Product Liability: The publisher can give no guarantee for information about drug dosage and application thereof contained in the book. In every individual case the respective user must check its accuracy by consulting other pharmaceutical literature.

Typesetting and printing: Konrad Triltsch GmbH, Graphischer Betrieb, 8700 Würzburg.
2125/3130-543210

Preface

This volume contains the proceedings of the symposium Pathochemical Markers in Major Psychoses, held in Vienna in July 1983.

The development of biological markers in psychiatric diseases, particularly in the field of neurochemistry, has made substantial progress during recent years although the multiple mechanisms of mental illness are still not fully understood. The greatest contribution has come through the development of new therapeutic agents that not only provide invaluable help for psychiatric patients but also serve as chemical tools for the investigation of the biological mechanisms underlying the disease. The catecholamine and serotonin hypotheses for major psychoses have been of particular heuristic value and have stimulated important research. However, the scope of our scientific endeavours has to be broadened to include other putative biological causes of psychoses, e.g., pathomorphological changes, aberrations in the metabolism of other amino acids and of lipids, or the formation of endogenous toxins.

This book presents new selected studies of the pathochemical bases of schizophrenia and affective psychoses. Although several topics were inevitably not included in this symposium, we nevertheless hope that it represents an integration of basic pathomorphological research with current clinical findings in the area of pathochemical markers in psychiatry.

October 1984

HELMUT BECKMANN
PETER RIEDERER

Contents

Contributors

MANFRED ACKENHEIL, M.D.
Psychiatrische Klinik der
Universität München
Nußbaumstr. 7
D-8000 München 2

MARGOT ALBUS, M.D.
Psychiatrische Klinik der
Universität München
Nußbaumstr. 7
D-8000 München 2

MIHÁLY ARATÓ, M.D.
Regional Neuropsychiatric Institute
P.O. Box 37
H-4321 Nagykálló

CSABA M. BANKI, M.D.
Regional Neuropsychiatric Institute
P.O. Box 37
H-4321 Nagykálló

HELMUT BECKMANN, M.D.
Zentralinstitut für Seelische Gesundheit
Postfach 59 70
D-6800 Mannheim

WADE BERRETTINI, M.D.
Section on Psychogenetics
Biological Psychiatry Branch
NIMH, Bethesda, M.D. 20205

HANS JÜRGEN BOCHNIK, M.D.
Klinikum der
Johann-Wolfgang-Goethe-Universität
Zentrum der Psychiatrie
Heinrich-Hoffmann-Straße 10
D-6000 Frankfurt 71

BRIGITTA BONEY, M.D.
Psychiatrische Klinik der
Universität München
Nußbaumstraße 7
D-8000 München 2

SUSAN M. BONHAM CARTER
Queen Charlotte's Hospital
Goldhawk Road
London W6 OXG

JACQUES BRUINVELS
Erasmus Universiteit Rotterdam
Faculteid der Geneeskunde
Dept. of Pharmacology
Postbus 1738
Rotterdam

LOTHAR DEMISCH, M.D.
Klinikum der
Johann-Wolfgang-Goethe-Universität
Zentrum der Psychiatrie
Heinrich-Hoffmann-Straße 10
D-6000 Frankfurt 71

EBERHARD GABRIEL, M.D.
Psychiatrisches Krankenhaus der Stadt Wien
Baumgartner Höhe 1
A-1145 Wien

THEO GASSER
Zentralinstitut für Seelische Gesundheit
Postfach 59 70
D-6800 Mannheim 1

Prof. Dr. WAGNER F. GATTAZ
R. Leoncio de Carvalho 253, apto 31
04003 São Paulo – SP – Brazil

ANNETTE GJERRIS, M.D.
Psychochemistry Institute
Righospitalet
9, Blegdamsvey
DK-2100 Copenhagen

KURT JELLINGER, M.D.
Krankenhaus der Stadt Wien-Lainz
Wolkersbergenstr. 1
A-1130 Wien

DAVID C. JIMERSON, M.D.
National Institute of Mental Health
Laboratory of Clinical Science
Bethesda, MD 20205

DANIEL P. VAN KAMMEN, M.D., Ph.D.
VA Medical Center
Highland Drive
Pittsburgh, PA 15206

X   Contributors

WELMOET B. VAN KAMMEN
VA Medical Center
Highland Drive
Pittsburgh, PA 15206

MARKKU LINNOILA, M.D.
VA Medical Center
Highland Drive
Pittsburgh, PA 15206

LEE S. MANN, M.D.
VA Medical Center
Highland Drive
Pittsburgh, PA 15206

EDWARD A. MUELLER, M.D.
Clinical Neuropharmacology Branch
National Institute of Mental Health
NIH – Building 10, Room 3D/41
Bethesda, Md., 20205

FRANZ MÜLLER-SPAHN, M.D.
Psychiatrische Klinik der
Universität München
Nußbaumstraße 7
D-8000 München 2

URSULA MÜNCH, M.D.
Psychiatrische Klinik der
Universität München
Nußbaumstraße 7
D-8000 München 2

DENNIS L. MURPHY, M.D.
Clinical Neuropharmacology Branch
National Institute of Mental Health
NIH – Building 10, Room 3D/41
Bethesda, MD., 20205

DIETER NABER, M.D.
Psychiatrische Klinik der
Universität München
Nußbaumstraße 7
D-8000 München 2

PHILIP T. NINAN, M.D.
VA Medical Center
Highland Drive
Pittsburgh, PA 15206

ZSUZSA PAPP, M.D.
Regional Neuropsychiatric Institute
P.O. Box 37
H-4321 Nagykálló

L. PEPPLINKHUIZEN, M.D., Ph.D.
Erasmus Universiteit Rotterdam
Faculteid der Geneeskunde
Dept. of Psychiatry
Postbus 1738
Rotterdam

OLE J. RAFAELSEN, M.D.
Psychochemistry Institute
Righospitalet
DK-2100 Copenhagen

FRANZISKA REINHUBER
Klinikum der
Johann-Wolfgang-Goethe-Universität
Zentrum der Psychiatrie
Heinrich-Hoffmann-Straße 10
D-6000 Frankfurt 71

GARVIN P. REYNOLDS, M.D.
MRC Neurochemical Pharm. Unit.
Brain Tissue Bank
Dept. of Neurol. Surgery and Neurology
Addenbrooke's Hospital
Cambridge, CB 2 2QQ

PETER RIEDERER, Ph.D.
Ludwig-Boltzmann Institut für klinische
Neurobiologie
Krankenhaus Wien-Lainz, Pav. XI
A-1130 Wien

MERTON SANDLER, M.D.
Queen Charlotte's Hospital
Goldhawk Road
London W6 OXG

MIKA SCHEININ, M.D.
VA Medical Center
Highland Drive
Pittsburgh, PA 15206

LARRY J. SIEVER, M.D.
Clinical Neuropharmacology Branch
National Institute of Mental Health
NIH – Building 10, Room 3D/41
Bethesda, Md., 20205

Neuromorphological Background of Pathochemical Studies in Major Psychoses

K. JELLINGER

The morphological substrates of major psychoses are controversial. While structural changes in the brain are commonly present in dementias and organic psychoses, no consistent morphological deficits have been demonstrated in functional psychoses that are often associated with a variety of pathochemical changes. Since the schizophrenia syndrome was originally identified by Kraepelin (1919), investigators have hypothesized that it could be due to specific brain abnormalities. In spite of the various modifications in the concept of schizophrenia, a tradition of searching for both functional and structural brain deficits has persisted.

Neuroradiological Approaches

Previous pneumencephalographic (PEG) and echoencephalographic studies indicating that chronic schizophrenia and "defect states" may be associated with brain atrophy have been confirmed by recent computer tomographic (CT) studies replicating the finding of increased ventricular size, sulcal enlargement (cortical atrophy), and atrophy of the anterior cerebellar vermis in some schizophrenics with "negative" symptoms and intellectual impairment (Table 1). Between 20% and 55% of all schizophrenics have been suggested to show some abnormalities in cerebral CT; findings in acutely ill patients are frequently normal, while up to 69% of patients with defect states show some ventricular and/or sulcal enlargement (Takashi et al. 1982; Gross et al. 1982; Andreasen et al. 1982; Weinberger et al. 1982; Dewan et al. 1983). In most studies the change is in the lateral ventricles, although in some series an increase in the third ventricular width was observed (see Weinberger et al. 1983). However, the extent of brain atrophy, not seen in all case series (Gluck et al. 1980; Jernigan et al. 1982; Boronow et al. 1983), varies among schizophrenic subtypes (Andreasen et al. 1982; Nasrallah et al. 1982; Gross et al. 1982).

Other neuroradiological data, such as occasional inversion of the hemispheral asymmetry (Luchins et al. 1979, 1982; Newlin et al. 1981; Naeser et al. 1981), reduced density of the cerebral tissue in the frontal area of the left (dominant) hemisphere (Golden et al. 1981), and diminution of frontal lobe cortical activity in terms of cerebral blood flow (CBF) (Ingvar and Franzen 1974; Ingvar 1980; Golden et al. 1981; Buchsbaum 1983), have been believed to indicate an "anatomical locus of the pathology of schizophrenia" that has been tentatively related to disorders of the subcorticocortical, thalamic, and mesolimbic projection systems (Scheller 1966; Stevens 1973; Ingvar 1980).

Table 1. Structural changes in schizophrenia – neuroradiological findings (see also Weinberger et al. 1983)

Author(s) (year)	Technique	n	Abnormality	Clinical state
Lempke (1935)	PEG	100	Enlarged ventricles	50% schizophrenics, 20% controls
Huber (1957, 1961)	PEG	260	Lat. + third ventricles	58.4% pathological defect states
Haug (1962)	PEG	101	Lateral ventricle	Chronic cases
Nagy (1963)	PEG	144	Lateral ventricle	80.8% pathological defects states
Storey (1966)	PEG	18	No differences from controls	
Asano (1967)	PEG	53	Enlarged ventricles	42% nuclear, 78% severe schizophrenics
Bliss (1976)	PEG	?	Lateral ventricle	Chronic schizophrenia
Feuerlein-Dilling (1967)	Echoenceph.			Defect states
Holden et al. (1973)	Echoenceph.	65	Third ventricle	Treatment resistance
Johnstone et al. (1976)	CT	18	Third ventricle	Intellectual impairment
Kingsley-Trimble (1978)	CT	?	Lateral ventricle	Chronic schizophrenia
Weinberger et al. (1979)	CT	20	Cerebellar atrophy	
Rieder et al. (1979)	CT	17	Lateral ventricle	Intellectual impairment
Gluck et al. (1980)	CT	?	No differences from controls	Chronic schizophrenia
Mundt et al. (1980)	CT	68	No differences from controls	Chronic schizophrenia
Donnely et al. (1980)	CT	15	Lateral ventricle	Intellectual impairment
Golden et al. (1980)	CT	42	Lateral ventricle	Intellectual impairment
Weinberger and Wyatt (1980)	CT	51	Lateral ventricle	Poor premorbid adjustment
Weinberger et al. (1980)	CT	20	Lateral ventricle	Treatment resistance
Golden et al. (1981)	CT	23	Lower density of left frontal area	Chronic schizophrenics
Tanaka et al. (1981)	CT	49	Lat. + third ventricles; cortical atrophy	Chronic patients
Takahashi et al. (1982)	CT	169	Third + lat. ventricles; cortical atrophy	Negative symptoms
Andreasen et al. (1982)	CT	52	Lateral ventricle (6%)	Negative symptoms
Gross et al. (1982)	CT	117	Third ventricle	Pure defect state −69% / Schizophrenia −28%
Heath et al. (1982)	CT	50	Cerebellar vermis	50% atrophy
Jernigan et al. (1982)	CT	30	No increased ventricles	Acute and chronic schizophrenia
Nasrallah et al. (1982)	CT	27	Lateral ventricle	Neg. symptoms, impaired cognition
Weinberger et al. (1982)	CT	52 (17 chronic)	Lateral ventricle	24% chronic cases
Crow et al. (1982b)	CT	?	Lateral ventricle	Defect states, involuntary movements
Nyback et al. (1982)	CT	41	Lat. + third ventricles	Acute and chronic schizophrenia
Boronow et al. (1983)	CT	23	No differences from controls	Chronic schizophrenia
Dewan et al. (1983)	CT	23	Third ventricle +	Chronic schizophrenia

Subsets of Schizophrenia and Morphology

Although after more than 100 years research into the neuropathology of schizophrenia has remained inconclusive (Peters 1967; Corsellis 1976; Jellinger 1980), there has been increasing evidence that structural brain deficits may be associated with some subsets of schizophrenia (Stevens 1983; Weinberger et al. 1983). Crow (1982 a, b) proposed that the disorder currently called schizophrenia may represent two overlapping syndromes: the acute type I syndrome of positive symptoms (hallucinations, delusions, thought disorders) with no evidence of intellectual impairment but good response to neuroleptics is believed to be a neurochemical disorder with an increase in numbers of dopamine D_2 receptors, while the chronic type II syndrome of negative symptoms (affective flattening, poverty of speech, loss of drive) with impairment of intellectual and cognitive functions, corresponding to the "defect state," may be a consequence of a degenerative process leading to cell loss in particular areas of the brain. It may well be, however, that the postulated genetically determined pathochemical processes do not necessarily lead to structural deficits of the brain that can be substantiated by current neuroradiological and/or neuropathological methods (Pearlson et al. 1981; Gross et al. 1982; Riederer and Jellinger 1982).

Two major problems will be critically discussed:

1. What structural changes of the brain in schizophrenia and other major psychoses have been hitherto reported and how can they be interpreted in terms of current neuropathology and neuroscience research?
2. What are the morphological factors to be considered in pathochemical studies of major psychoses in human postmortem brain?

Although a large variety of structural changes have been reported in both acute and chronic schizophrenia (see Tables 2–5), most investigators emphasize that to date no controlled study has been able to demonstrate any consistent gross or microscopic abnormality in schizophrenic brains that might be characteristic of this or any other type of psychosis (Corsellis 1976; Peters 1974; Riederer and Jellinger 1982). Neuropathological findings after neuroleptic long-term therapy and in patients with dyskinesia will not be discussed (see Christensen et al. 1970; Jellinger 1977).

Neuropathology of Catatonia

In acute psychoses with lethal outcome, including catatonia, frequent autopsy findings are cerebral edema with narrow ventricles, signs of increased intracranial pressure, vascular congestion with small hemorrhages, and occasional secondary anoxic brain lesions. These findings are considered to result from disorders of electrolyte metabolism, hypoxia, and vascular disorders, while occasional inflammatory CNS lesions are probably caused by coincidental viral infections causing the clinical picture of acute psychosis (Table 2). Most of the reported cytological changes, e.g.,

Table 2. Neuropathology of acute catatonia

Author(s) (years)	Morphological CNS findings	Comments
Buscaino (1920)	Neuronal loss in globus pallidus	Secondary anoxic lesions
Stauder (1934)	Cerebral edema	
Malamud and Boyd (1939)	Anoxic cortical lesions	Secondary lesions of anoxic or vascular origin
Shulack (1945)	Encephalic congestion; swelling of cortical and subcortical neurons; small petecchial hemorrhages	Nonspecific lesions
Adland (1947)	Perivascular lymphocytic infiltrates	
Vogt and Vogt (1952)	Clusters of neurons with vacuolation and fatty degeneration in dorsomedial thalamus, anterior cingulate gyrus, 3rd temporal gyrys, globus pallidus	1. Autolytic changes (Heyck 1954)
Buttlar-Brentano (1952)	Dwarf cells in nucleus basalis, ventral pallidum, nucleustria terminalis, periventricular, supraoptic, tuberal and mammillary nuclei	2. Nonspecific (Hassler 1967; Peters 1967)
Hopf (1952)	Dwarf cells and lipid degeneration in medial and dorsomedial thalamus	
Scharenberg and Brown (1954)	Pathological changes of nerve fibers	Silver staining
Hempel and Treff (1959)	30%–40% neuronal loss in mediodorsal thalamus	
Penn et al. (1972)	Patchy loss of neurons in insular cortex; perivascular lymphocytic infiltrates in hippocampal gyrus; slight focal lymphocytic meningitis	Viral infection (?)
Dom and DeSaedeleer (1981)	Decreased diameter of microneurons in striatum; slight atrophy of medial thalamus; 50% decrease of Golgi II microneurons in posterior thalamus	Hypotrophy of neostrial inhibitory Golgi II neurons (?)
Stevens (1982)	Fibrillary gliosis in basal forebrain, amygdala, periventricular thalamus, hypothalamus, limbic system; neuronal loss in globus pallidus	Postinflammatory changes (?)

neuronal loss, lipid degeneration, vacuolation, and dwarf cells (*Schwundzellerkrankung*) in subcortical nuclei have not been confirmed in control studies (Heyck 1954) and are either claimed as nonspecific or postmortem (autolytic) changes, or related to electroconvulsive treatment or to agonal and extracranial disease (see Hassler 1967; Peters 1967, 1974; Corsellis 1976).

Neuropathology of Chronic Schizophrenia

The gross and histopathological findings reported in dementia praecox and schizophrenia are summarized in Tables 3–5.

Table 3. Neuropathology of dementia praecox and schizophrenia – macroscopic findings

Author(s) (year)	Material	Autopsy findings
Witte (1942)	500 autopsies 281 males 209 females	Brain weight normal for age; mean 1375 g (males), 1210 g (females)
Peters (1937)	Autopsies	Brain weight normal for age
Broser (1949)	219 autopsies, chronic cases (mean duration 19.8 yrs)	Normal brain weight, no internal hydrocephalus
Wildi et al. (1967)	75 autopsies, chronic schizophrenics	Cerebral atrophy; atherosclerosis, vascular hyalinosis, senile and vascular cerebral lesions rarer than in controls
Rosenthal and Bigelow (1972)	Autopsies, chronic schizophrenics	Significantly wider corpus callosum; normal brain weight, cortical thickness, and volume of thalamus and temporal lobe
Heath et al. (1979, 1982)	CAT, surgery, autopsy	Atrophy of upper cerebellar vermis
Jellinger (1980)	93 autopsies, chronic schizophrenia	13% cerebral atrophy and/or inter- nal hydrocephalus more pro- nounced than in age-matched controls
Crow et al. (1982)	Chronic type II autopsies	Wider ventricles, decreased brain weight, decreased volume of caudate nucleus and cortex (compared to affective disorders)

Table 4. Neuropathology of dementia praecox and schizophrenia – cerebral biopsy findings

Author(s) (year)	Biopsy findings	Comments
Elvidge and Reed (1938)	Swelling of oligodendroglia in white matter	Fixation artefacts
Kirschbaum and Heilbrunn (1944)	Degeneration of neurons and glia in 10/11 cases	Nonspecific data; no controls
Hartelius (1948)	Swelling of shrinkage of cortical neurons, decreased nucleoprotein contents	
Hyden (1952)	Decreased nucleoproteins in cortical neurons	
Meyer (1952)	Nothing abnormal	
Palma and Sotelo (1952)	Cortical atrophy	Macroscopic findings at frontal lobotomy
Miyakawa et al. (1972)	Hyperplasia of smoth endoplasmic re- ticulum in neurons and oligodendroglia; abnormal synaptic membranes; granular deposits in cytoplasm of oligodendroglia	Suggests disorders of axonal flow, in- creased protein syn- thesis

Table 5. Neuropathology of dementia praecox and schizophrenia – histological findings

Author(s) (year)	Histological findings	Comments
Alzheimer (1897)	Pallor, loss of cortical pyramidal cells	Nonspecific findings
Kraepelin (1919)	Lipid accumulation and atrophy of cortical neurons	
Buscaino (1920)	Disintegrating plaques in white matter; formation of metachromatic globules	Fixation artefacts (Lampert 1972)
Nagasaka (1925)	Loss of large striatal neurons; depigmentation of ⅔ of cells in substantia nigra	
Münzer (1926)	Shrinkage and lipid accumulation in neurons of cortex and Ammonshorn; focal cell loss in cortical layers III–V; neuronophagias and diffuse gliosis	Nonspecific, partly artefacts
Josephy (1923)	Sclerosis and lipid accumulation in pyramid neurons, layer II prefrontal cortex and hippocampus; calcification in the pallidum, neuroglial nodules in striatum, thalamus, and brain stem	Nonspecific
Spielmeyer (1931) Hechst (1931)	Lipid degeneration in the neurons of the 3rd cortical layer	Nonspecific
Dide (1934)	Hypertrophy, dystrophy of hypothalamic neurons, glial plaques	
Morgan and Gregory (1935)	Degenerative changes and gliosis in hypothalamus	Nonspecific
Fünfgeld (1937) Miskolczy (1937)	Focal loss of cortical neurons (*Lückenfelder*) in layers III and V; lipid degeneration in medial thalamic nuclei, temporal gyrus, and cingulate gyrus	Normal cytoarchitectonic variants
Meyer (1952)	Nonspecific cortical changes and marginal gliosis	Nonspecific findings attributed to electroconvulsive therapy
Wahren (1952)	Neuronal degeneration in hypothalamus	
Bruetsch (1952)	Rheumatic nodules and endothelial proliferation	(9 of 100 cases)
Van der Horst (1952)	Degeneration of neurons; fibrillary gliosis; small hemorrhages; perivascular infiltrates	patients with rheumatic endocarditis
Bäumer (1952)	Neuronal loss in mediodorsal thalamus	
Palma and Sotelo (1952)	Pathological changes in nerve fibers	Silver stainings; increased axonal transport and disorders of protein synthesis (?)
Tatetsu (1964)	Thickened apical dendrites and axons in areas 4, 7, 8, 10, 11, 17, 18 (prefrontal cortex)	
Miyakawa (1964)	Thickened apical dendrites and axons in limbic system	
Glezer and Soukhoroukova (1966)	Degenerative changes in neurons and astroglia	

Table 5. (continued)

Author(s) (year)	Histological findings	Comments
Russkych (1969)	Neuronal loss and demyelination in frontal lobe, reticular thalamus, thalamocortical, hypo-thalamus, and limbic systems	Anoxic changes
Nieto and Escobar (1972)	Gliosis in reticular formation, hypothalamus, septum nuclei, medial, anterior thalamus, peri-aqueductal gray, hippocampus	4/10 cases
Colon (1972)	Reduction of cortical thickness by 20%; cortical neuronal loss of 55% with accentuation in deep cortical layers (70%)	Demented schizophrenics; 45% neuronal loss in nondemented senile brains
Peters (1974)	Swelling of apical dendrites and axons in cortical area 10	
Fishman (1975)	Glial nodules in brain stem, reticular nuclei, trigeminal nuclei	6/8 cases; resembling herpes encephalitis
Heath et al. (1979)	Slight atrophy of cerebellar vermis without histological changes	
Weinberger et al. (1980)	Atrophy of cerebellar vermis	3/12 schizophrenics; 1/37 controls
Scheibel and Kovelman (1980)	Disarray of hippocampal pyramidal cells	10 schizophrenics
Averback (1981)	Neuronal loss in nucleus basalis of Meynert (subnucleus ansae peduncularis)	Young patients (13 cases)
Hankhoff and Peress (1981)	Microglial nodules and perivas-cular infiltrates in brain stem	1/8 schizophrenics; 7/27 controls
Stevens (1982)	Fibrillary gliosis in periventric-ular, periaqueductal gray of mesencephalon, basal forebrain, limbic system, hypothalamus, midbrain tegmentum, etc. Neuronal loss in globus pallidus, nucleus accumbens, nucleus basalis	3/4 of autopsy cases; inflammatory residuals in some cases
Bogerts et al. (1982)	Significant decrease in volume of nigrostriatal areas, reduction of mean volume of glial nuclei, reduction of mean volume of neurons in mesolimbic system	6 cases Vogt's material
Arendt et al. (1983)	Normal neuronal counts and den-sity in nucleus basalis and globus pallidus	Catatonia
Jellinger (1983) (unpublished results)	Normal neuronal counts, maxi-mum and mean cell densities in nucleus basalis of Meynert	Chronic schizophrenics
Nasrallah et al. (1983)	Gliosis in corpus callosum	

Macroscopic Changes

Gross examination of schizophrenic brains either failed to find differences in weight and size as compared to age-matched mentally normal controls or showed decreased brain weight, wider ventricles, slight diffuse cortical atrophy, or wider corpus callosum in some cases (Table 3). Crow et al. (1982) examined the brains of over 200 psychiatric patients without the benefit of a normal control series. When the patients were matched for age, schizophrenics had larger ventricles on coronal sections and lower brain weight than depressed patients; those with intellectual impairment (average age 70.6 years) had larger ventricles and decreased volume of the striatum postmortem than those without dementia (average age 54 years).

Histopathology

The majority of cytological and other histopathological changes in the cerebral cortex and white matter described in brain biopsies (Table 4) and postmortem materials from schizophrenics (Table 5) have also been observed in nonpsychiatric brains, and have therefore, been explained by either coincidental or preparatory factors (Meyer 1952). Focal neuronal "loss" in the cortical layers III and V (*Lückenfelder*), regarded by earlier investigators as "specific" changes in schizophrenia (Alzheimer 1897; Josephy 1923; Fünfgeld 1937), have also been observed in nonpsychiatric controls and are regarded as variants in cortical cytoarchitecture (Spielmeyer 1930; Peters 1967). According to Colon (1972), neuronal loss in cerebral cortex, frequently observed in chronic schizophrenia with mental impairment, amounts to 57% with accentuation in the deep cortical layers, and is higher than in aged nondemented patients (45%) but similar to that in senile dementia (55%). These neuronal losses are regarded as a morphological correlate of dementia in defect states, while atrophy of the upper cerebellar vermis (Heath et al. 1979; Weinberger et al. 1980) has been observed only in a small number of schizophrenic brains (Table 8).

Thickened apical dendrites and axons in cerebral cortex and limbic areas (Tatetsu 1964, 1974) have been confirmed by cerebral biopsy findings demonstrating hyperplasia of the smooth endoplasmic reticulum in neurons and oligodendroglia, increase of tubular structures in nonmyelinated axons, membrane-bound lysosomes in postsynaptic areas, abnormal synaptic membranes, and granular deposits in the cytoplasm of neurons and oligodendroglia (Miyakawa et al. 1972). These findings, some of which resemble the ultrastructural changes in "axonal dystrophies" (Jellinger 1973), suggest some disorders of axonal flow, synaptic remodeling processes, and disorders of protein synthesis in neurons and oligodendroglia.

Inflammatory nodules and occlusive endarteritis of small cerebral arteries observed in some schizophrenic brains (Bruetsch 1952) are the consequence of rheumatic valvular heart disease and give no evidence of rheumatic pathology of schizophrenia (Table 5). Occasional glial nodules and fibrillary gliosis in the brain stem, regarded as sequelae or residuals of inflammatory processes, encephalitis-like disorders, or chronic viral infections (Fishman 1975; Stevens 1973, 1982), are also to be encountered incidentally in nonpsychiatric autopsies and other disease processes, and are hence explainable as nonspecific or in terms of associated CNS diseases that do not appear related to primary psychoses (Corsellis 1976; Jellinger 1980; Hankhoff and Peress 1981). On the other hand, there is no evidence so far to support the hypothesis of some viral agent either causing a primary neurochemical disorder in susceptible individuals or inducing a more widespread encephalitis-like disorder, leading to neuronal degeneration and structural changes in the brain of schizophrenics (Crow 1983). Whether the demonstration of viral antigen of CMV (cytomegalovirus) or varizella-zoster in the neurons of the olivary nucleus, thalamus, and hypothalamus in four of six schizophrenics and in four of twelve control brains is in any way related to the etiology of schizophrenia remains to be determined (Stevens et al. 1983).

This also holds true for the fibrillary gliosis in the diencephalon, periaqueductal region of the brain stem, basal forebrain, and mesolimbic areas observed in 40% of the cases of Nieto and Escobar (1972) and in three-quarters of the schizophrenic brains reported by Stevens (1982). These data, however, have not been confirmed by careful histopathological studies by others, including Corsellis (1983, personal communication), who examined most of the psychiatric autopsy material presented by Crow et al. (1982) and who found such extended fibrillary gliosis only in one single case of schizophrenia. In our recent series of 95 autopsy cases of schizophrenia we observed this kind of fibrillary gliosis in the basal forebrain and brain stem only in six instances (6.3%), while another three cases showed slight cerebellar cortical atrophy (Table 8).

Neuronal vacuolation, lipid degeneration, and dwarf cells (*Schwundzellerkrankung*) in several subcortical nuclei, particularly in the mediodorsal thalamus, and nucleus basalis, regarded by Vogt and Vogt (1952), Buttlar-Brentano (1952), and Averback (1981) as the results of premature cell ageing or genetically determined hypoplasia, have not been confirmed by later investigators, and have been explained by agonal and preparatory artefacts (Heyck 1954; Hassler 1967; Peters 1967; Corsellis 1976). On the other hand, neuronal losses in the medial and posterior thalamus (Hopf 1952; Hempel and Treff 1959) have been confirmed by recent cytometric analyses in the striatum, posterior thalamus (Dom and de Saedeleer 1981), and mesolimbic system (Bogerts et al. 1982), and have been interpreted as hypoplasia or deficit of inhibitory Golgi II microneurons which could be hypersensitive to dopamine. Recently, Bogerts et al. (1983) reported a 20% reduction of the volume of the lateral two-thirds of the substantia nigra without loss of nerve and glial cells, indicating a reduction of the neuropil-containing striatonigral fibers. However, neuronal loss in the globus pallidus and substantia innominata reported by Hopf (1952), Buttlar-Brentano (1952), Averback ((1981), and Stevens (1982) have not been confirmed by recent morphometric studies in schizophrenic brains (Arendt et al. 1983). The cholinergic forebrain system, particularly the nucleus basalis of Meynert of the substantia innominata, representing the main cholinergic input to the cerebral neocortex and limbic and related structures, appears to play an important role in cognitive functions and behavioral reactions (Doyle et al. 1983). In both Alzheimer's disease and senile dementia of the Alzheimer type, compelling evidence has been developed that there is a selective degeneration of acetylcholine-releasing neurons in the basal forebrain with neuronal loss in up to 65% of controls, causing a

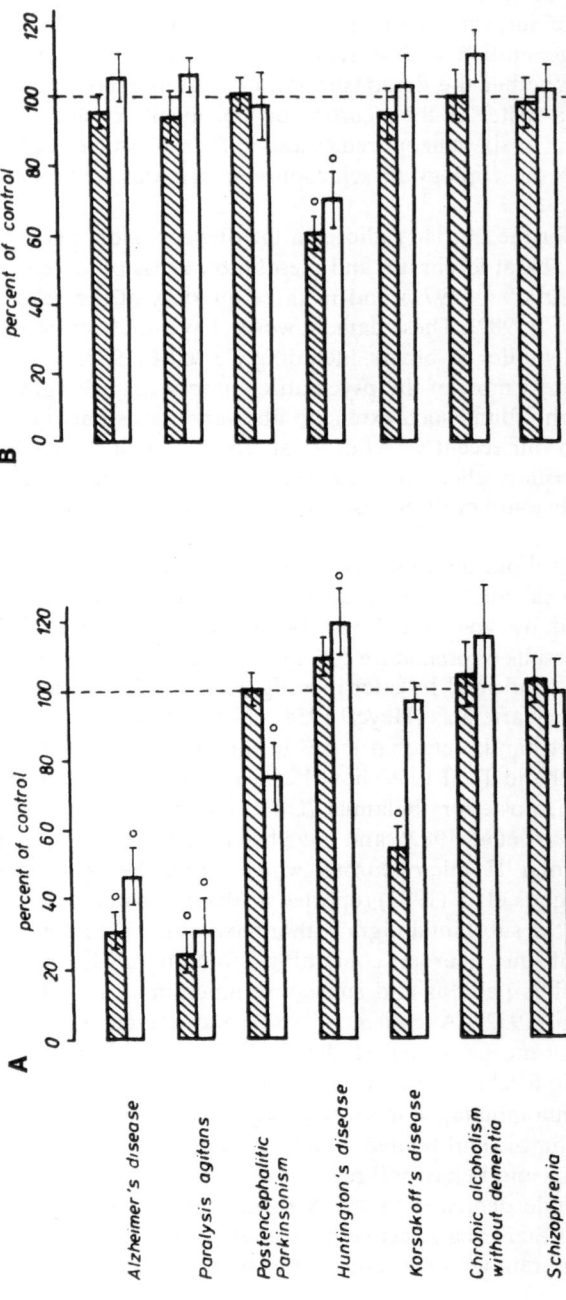

Fig. 1A, B. Neuronal counts (*hatched bars*) and maximum population density (*white bars*) **A** in the nucleus basalis of Meynert (nbM) and **B** in the external segment of the globus pallidus in various neuropsychiatric disorders compared to age-matched controls. Severe neuronal loss in nbM is seen in Alzheimer's and Parkinson's disease and in Korsakoff's syndrome, considerable neuronal loss in the globus pallidus in Huntington's chorea. No neuronal loss in either nucleus is observed in schizophrenia. *Bars with circle* indicate significant difference from the controls (Mann-Whitney U-test, $p < 0.01$). (After Arendt et al. 1983)

disorder of cortical cholinergic innervation which has been related to deterioration of memory and cognition (Doyle et al. 1983; Candy et al. 1983). Similar loss of cholinergic neurons in the nucleus basalis has recently been observed in Parkinson's disease, most accentuated in demented Parkinsonian patients showing neuronal loss of up to 70% of control values (Arendt et al. 1983; Candy et al. 1983; Whitehouse et al. 1983; Riederer and Jellinger 1983). Although there is some evidence of a deficit of choline acetyltransferase, the biosynthetic enzyme for acetylcholine, in the septohippocampal complex in schizophrenia (Bird et al. 1979), from recent morphometric studies by Arendt et al. (1983) and our own experience in a small series of chronic and defect states there is no evidence for a substantial reduction of neurons in the basal forebrain in schizophrenia (Fig. 1).

Similar reservations are necessary with regard to frequently postulated structural lesions in the nucleus accumbens and substantia perforata anterior (Stevens 1982), in which increased levels of dopamine have been demonstrated in schizophrenia (Farley et al. 1978; Crow et al. 1982). However, neither Corsellis (1983, personal communication) and other experienced neuropathologists nor we ourselves were able to find consistent structural changes in these nuclei.

In conclusion, none of the postulated structural changes reported in schizophrenic brains appear to be consistent or typical for this or other major psychoses, although it seems likely that some structural changes, including dilatation of the external and internal CSF spaces and neuronal loss in some brain areas, do occur in chronic schizophrenics and defect states. They are, however, to be regarded rather as the morphological substrates of mental impairment and organic psychosyndromes in these patients than as the basic disease process. In view of recent morphometric and biochemical studies, the morphology of some subcortical areas, particularly of the limbic, striatonigral, and forebrain systems, in schizophrenic brains needs further investigation.

Organic CNS Disorders ("Symptomatic" Schizophrenia)

There are many reports on schizophrenia-like psychoses which have occurred in association with organic disorders affecting the nervous system (Davison and Bagley 1969), and it is well known that a variety of organic brain diseases may mimic the symptoms of schizophrenia (see Peters 1967, 1974; Malamud 1975; Corsellis 1976; Jellinger 1980). There has been a considerable degree of unexpected cerebral pathology in various samples from schizophrenic patients examined by CT or at autopsy. The incidence of organic brain lesions in series of patients with the clinical diagnosis of schizophrenia ranges from 9% in clinical and CT series (Owens et al. 1980) to 31% in autopsy series (Table 6). The neuropathology findings in schizophrenia-like psychoses include acute and chronic inflammatory lesions, e.g., herpes simplex encephalitis, epidemic encephalitis, rheumatoid lesions, subacute sclerosing panencephalitis, and Creutzfeldt-Jakob disease; demyelinating diseases, such as diffuse, concentric, or multiple sclerosis, posttraumatic and postanoxic encephalopathies, residual states of intoxication, vascular lesions, and brain infarction; metabolic diseases, e.g., neurolipidoses, metachromatic and other leukodystrophies;

Table 6. Organic brain disease in patients with clinical diagnosis of schizophrenia

Author(s) (year)	Material		Brain lesions
Owens et al. (1981)	136 patients CT scans	9%	(cerebral infarctions, subdural hematomas, intracranial tumors, etc.)
Alsen (1978)	280 autopsies	21%	(Alzheimer's disease, SDAT, subdural hematomas, brain infarction, tumors, storage diseases)
Jellinger (1980)	101 autopsies	31%	(Alzheimer's disease, SDAT, vascular lesions, tumor, demyelinating disorders, cerebral angiomas)
Jellinger (1983)	100 autopsies	27.0%	(various organic brain disorders and lesions)
		17.0%	(superimposed or secondary cerebral lesions)
		44.0%	organic brain lesions

Table 7. Neuropathology findings in schizophrenia (unselected autopsy material)

Clin. diagn./ morphology	Schizo-phrenia	Paranoid schizo-phrenia	Defect state	Schizo-phr. and epilepsy	Cata-tonia	Hebe-phrenia	Total
Normal brain	9	8	5	–	1	1	24
Senile brain	12	7	13	–	–	–	32
Brain edema	5	3	2	–	3	1	14
Hydrocephalus	2	1	3	–	–	–	6
Alzheimer's disease	1	3	2	–	–	–	6
Vascular encephalitis; cerebral infarction	6	1	1	–	–	–	8
Brain tumor[a]	2	2	2	–	–	–	6
Demyelinating disease[b]	1	1	–	–	–	1	3
Old head injury[c]	–	–	(2)	–	–	–	(2)
Arteriovenous angioma	–	–	–	2	–	–	2
Lobotomy scars[c]	(1)	(1)	–	–	–	–	(2)
Total	38	26	28	2	4	3	101

[a] Two meningiomas, 1 ganglioglioma of hypothalamus, 1 lipoma of midbrain tegmentum
[b] One acute disseminated encephalomyelitis, 1 multiple sclerosis, 1 metachromatic leukodystrophy
[c] Not included: head injury after onset of disease, lobotomy (therapeutic)

Table 8. Neuropathology findings in major psychoses (autopsy material from Vienna Psychiatric Hospital 1978–1983)

Diagnosis (ICD no.)	Paranoid schizophr. (295.3)	Acute/ other schizophr. (295.0/4)	Schizophr. defect (295.6)	Schizoaff. psychosis (295.7)	Depress. M.D.K. (296.1, 3)	Paranoid psychosis (297.1)	Senile paranoid (293.0, 290.1)	Unclassified	Total n	Total %	Median age at death
Normal/superimposed lesion	16	5	33	5	4	6	1	3	73	73.0	
Normal	12	3	16	4	4	5	1	2	47	47.0	63.6
Brain stem gliosis	–	1	5	1	–	1	–	–	6	4.0	70.0
Cerebellar atrophy	1	–	–	–	–	–	–	–	3	3.0	64.5
SDAT	–	–	8	–	–	–	–	–	8	8.0	79.2
Anoxia, Hemorrhage	2	–	–	–	–	–	–	1	3	3.0	58.0
Metastases	1	–	1	–	–	–	–	–	2	2.0	62.5
Hepat. encephalop.	–	1	3	–	–	–	–	–	3	3.0	52.0
CNS leukemia	–	–	–	–	–	–	–	–	1	1.0	50.0
Other process	6	0	5	1	1	10	4	0	27	27.0	
SDAT	2	–	–	–	–	4	2	–	7	7.0	81.0
Alzheimer's disease	1	–	–	–	–	2	1	–	4	4.0	67.0
Senile vasc. enceph.	1	–	3	–	1	–	1	–	6	6.0	81.7
Vasc. encephalop.	–	–	1	–	–	1	–	–	2	2.0	70.3
Alcohol. enceph.	–	–	1	1	–	–	–	–	1	1.0	65.0
Meningioma	1	–	–	–	–	–	–	–	2	2.0	61.0
Multiple sclerosis	1	–	–	–	–	–	–	–	1	1.0	74.0
Fahr's disease	–	–	–	–	–	3	–	–	3	3.0	70.0
Total	22	5	38	6	5	16	5	3	100	100.0	
Median age	62.3	57.0	67.1	56.5	68.4	76.1	77.0	48.0			

ICD, *International Classification of Diseases*, 9th edition; M.D.K., manic-depressive syndrome senile and presenile atrophies, including Pick's and Alzheimer's disease and Huntington's chorea; and intracranial tumors and angiomas, particularly those affecting the frontal and temporal lobes, basal ganglia, diencephalon, and limbic system (Malamud 1967; Peters 1974). In an unselected series of 101 autopsy cases with the clinical diagnosis of schizophrenia from a psychiatric center in the pre-CT era, only 24% had normal brains; 32% had senile atrophy, while 31% showed a variety of organic disorders (Table 7). In another autopsy series of 100 cases with major psychoses, 27% showed various organic brain lesions (Table 8). The rather frequent presence of – clinically often unexpected – organic disorders affecting the CNS or of other pathological brain lesions is an important factor to be considered in biochemical studies of major psychoses, and the presence of such organic brain deficits should be excluded by careful neuropathological examination prior to pathobiochemical studies.

Limiting Factors of Pathobiochemical Studies

When chemical studies in search of biochemical markers in major psychoses are being performed, a series of factors should be considered:

Clinical information, including diagnosis, classification, degree of dementia, drug treatment, and associated diseases
Circumstances and major causes of death, duration of agony, general autopsy findings, and postmortem time
Gross and histopathological findings on the CNS (in order to exclude the presence of organic brain lesions before forwarding the tissue to the neurochemist).

The importance of neuropathological examination prior to neurochemical studies can be demonstrated in two personal autopsy series:

1. Our Schizophrenia Research Program, a consecutive 5-year autopsy series including schizophrenia cases and cases of other major psychoses, performed at the Vienna Psychiatric Hospital, included 100 cases which underwent careful gross and microscopic examination using routine methods. Neuropathological investigation revealed nothing abnormal in 47.0%, while 9.0% showed either brain stem gliosis or mild cerebellar atrophy, occasionally reported in chronic schizophrenia (Stevens 1982; Heath et al. 1982; Weinberger et al. 1983), and another 17% showed superimposed acute or age-related changes of the brain (Table 8). In 27.0% there were signs of other organic brain disorder, e.g., senile and presenile atrophy, vascular, alcoholic, or demyelinating lesions, and brain tumors (meningiomas), while three cases showed abortive Fahr's disease, the relationship of which to schizophrenia is unknown. While the age at death did not vary considerably between patients with paranoid schizophrenia and those with defect states in our series, the schizophrenic patients without pathological brain deficits were older (63.6 years) than in most previous autopsy series (Vogt and Vogt 1952; Crow et al. 1982; Stevens 1982). Patients with paranoid psychoses (76–77 years) were older than the average schizophrenic, and often showed senile brain atrophy. Schizophrenic patients with primary or superimposed senile atrophy (SDAT) or senile vascular lesions were significantly

older (79.2–81.7 years) than schizophrenics without organic brain deficits (Table 8). Although the relationship of both primary and superimposed organic brain deficits to the pathophysiology and the pathochemical markers of schizophrenia is unclear, all cases with primary and secondary (superimposed) organic lesions were excluded from our pathobiochemical studies.

2. Similar strict regulations were followed for control cases. Among 117 controls without clinical evidence of neuropsychiatric disorders submitted to our brain bank in 1980–1981, only 58.2% showed nothing abnormal at neurohistological examination, while all the others revealed senile, vascular, anoxic, or other structural deficits, and were therefore regarded as unsuitable as "normal" controls for neurochemical analysis (Table 9).

With regard to classification, the correlations between Feighner's Research Criteria (FRC) (Feighner et al. 1972) and neuropathological findings in our autopsy series of cases with major psychoses appear to be of interest. Among the 100 cases the clinical histories of 74% fulfilled FRC of schizophrenia, while 26% did not. Among the Feighner-positive cases, 59.4% had normal brain morphology and were forwarded to pathobiochemical analysis, while among those which did not fulfil FRC only 35% showed no structural abnormalities of the brain (Table 10). All the others showed either superimposed or secondary CNS lesions or other basic CNS disorders, and therefore had to be excluded.

In addition to clinical, therapeutic, and neuropathological data, information on the antemortem and agonal states and major causes of death as revealed by general autopsy is essential for the validity of biochemical measurements in human postmortem brain specimens, as shown by extensive neurochemical correlative studies (Bowen et al. 1976, 1977; Spokes 1979; Perry and Perry 1983). In our autopsy series of cases with major psychoses, about 23% each died from pulmonary embolism and pneumonia and 5% from aspiration and acute bolus, all preventing long antemortem cerebral hypoxia (Table 11). In patients dying from cardiac failure and pulmonary edema (13%) the influence of terminal hypoxia on CNS neurotransmitters and receptor kinetics cannot be definitely excluded. Patients dying from hepatic coma, uremia, severe gastrointestinal hemorrhage, acute leukemia, and endocarditis were automatically excluded from neurochemical analysis, since the influence of these conditions on pathochemical markers is well established.

Table 9. Comparison of clinical diagnosis and neuropathology findings in 117 controls of Vienna brain bank (1980–1981)

Clinical diagnosis	n	%	Neuropathology	n	%
Myocardial infarction	17	14.5	Nothing abnormal	68	58.2
Heart failure	7	6.0	(Sen) Vasc. encephal.	14	12.0
Pneumonia/embolism	22	18.8	SDAT	9	7.7
Carcinoma	32	27.4	Anoxia	10	8.5
Diabetes	5	4.3	Brain infarction	8	6.8
Arteriosclerosis	11	9.4	Metastasis	2	1.7
Plasmocytoma	6	5.1	Others	6	5.1
Others	17	14.5		117	100.0
	117	100.0			

Table 10. Schizophrenia and paranoid psychoses: correlations between Feighner's Research Criteria and neuropathology

	Positive	Negative	Total
Normal brain	44	9	53
SDAT	10	6	16
Alzheimer's disease	2	2	4
Senile vascular encephalopathy	4	2	6
Vascular encephalopathy	1	1	2
Anoxia (old, recent)	2	1	3
Tumors			
Meningioma	1	1	2
Metastases	2	–	2
Hepatic encephalopathy	2	1	3
Glial nodules in brain stem	–	1[a]	1[a]
(trigem. nucl.)			
Multiple sclerosis	1	–	1
Fahr's disease	2	1	3
Alcoholic encephalopathy	–	1	1
Cerebellar atrophy	2	1	3
CNS leukemia	1	–	1
Total	74	27	101

[a] Combined CNS lesions

Table 11. Causes of death in autopsy series of schizophrenia (Vienna Psychiatric Hospital 1978–1982)

Cause of death	n
Pulmonary embolism	23
Pneumonia	23
Cardiac insuff., pulmonary edema	13
Metastasizing carcinomas	10
Peritonitis, ileus	4
Myocardial infarction	3
Hepatic coma (liver cirrhosis, hepatic carcinoma)	3
GI hemorrhages (gastric ulcer, etc.)	4
Aspiration, bolus	6
Pyelonephritis, uremia	4
Acute leukemia	2
Endocarditis	2
Unknown (no autopsy)	3
Total	100

GI, gastrointestinal

In conclusion, the fact that neuropathology demonstrated structural deficits in the brain due to primary, secondary, or superimposed organic CNS disorders in 45% of an autopsy sample of patients with the clinical diagnosis of major psychosis, particularly schizophrenia, and in 42% of a series of "normal" controls without clinical evidence of neuropsychiatric disorders underlines the importance of carrying out exact pathomorphological examination of the brain before performing neurochemical studies. The frequent occurrence of unexpected structural changes in human brains, as well as the large variety of antemortem and agonal conditions possibly influencing pathochemical markers in the brain and other variables, may explain some discrepancies in recently published data on human postmortem brain pathochemistry of major psychoses.

The goals for future neuropathological research in major psychosis are twofold:

1. Past histological studies attempted to find a single neuropathology common to all schizophrenics, but to data no study has been able to demonstrate any consistent structural CNS abnormality in schizophrenia. Two factors may account for this. First, some areas of potential interest in the brain, particularly subcortical regions, have not been systematically examined in well-controlled neuropathological studies, but some of them, e.g., the nucleus basalis, are now being increasingly investigated by combined morphometric and neurochemical studies. Second, the evidence from clinical, CT, and neurochemical investigations is that there are several subgroups of schizophrenia, some of which show a variety of brain abnormalities neuroradiologically, and some of which show none. Neuropathological strategies in the future would be to attempt to identify structural deficits of the brain specific to each of these separate subgroups.

2. The other important task of neuropathology is to exclude unexpected and unrelated structural changes of the brain in major psychoses in order to provide appropriate brain material for neurochemical investigation.

Summary

The morphological substrates of major psychoses are controversial. While structural changes in the brain are commonly present in dementias and organic psychoses, no consistent morphological abnormalities have been substantiated in schizophrenia, although both neuroradiological (computer tomographic, CT) and neuropathological studies suggest that structural deficits including brain atrophy and neuronal loss or gliosis in some brain areas may occur in some cases of chronic schizophrenia and defect states. The majority of these changes, however, are regarded as the substrates of intellectual impairment rather than of the primary psychosis, while a wide variety of structural and cytological changes described in schizophrenic brains are explained as nonspecific or due to coincidental or agonal changes that do not appear to be related to primary psychoses. More recent morphometric analyses did not confirm previously reported deficits in some subcortical areas (nucleus basalis, globus pallidus) of schizophrenic brains that had been tentatively related to some pathochemical changes or chronic infections leading to neuronal degeneration. On the other hand, the rather frequent presence of unexpected organic lesions affecting the brain in schizophrenic patients and a variety of other limiting factors (lethal disease, agonal state, etc.) must be considered in pathochemical studies of human postmortem brain. Neurohistology in a consecutive autopsy series of 100 cases of schizophrenia and other major psychoses revealed nothing abnormal only in 47.0%; 9.0% had brain stem gliosis or mild cerebellar atrophy (occasionally seen in schizophrenia), while the others showed acute lesions superimposed on the basic process (17%) or a variety of organic brain disorders (27.0%) including senile, vascular, and inflammatory lesions or tumors. The fact that structural deficits probably unrelated to the primary disease process are demonstrated in 31%–45% of the brains in major psychoses may explain some of the discrepancies in recently published data in human postmortem brain pathochemistry and emphasize the importance of carrying out exact morphological analyses before performing biochemical studies.

References

Adland ML (1947) Review, case studies, therapy, and interpretation of the acute exhaustive psychoses. Psychiatr Q 21:39–69
Alzheimer A (1897) Beiträge zur pathologischen Anatomie der Hirnrinde und zur anatomischen Grundlage der Psychosen. Mschr Psychiat Neurol 2:82–120
Andreasen NC, Olsen SA, Dennert JW, Smith MR (1982) Ventricular enlargement in schizophrenia. Am J Psychiatry 139:297–302
Arendt T, Bigl Y, Arendt A, Tennstedt A (1983) Loss of neurons in the nucleus basalis of Meynert in Alzheimer's disease, paralysis agitans and Korsakoff's disease. Acta Neuropathol (Berl) 61:101–108
Asano N (1967) Pneumoencephalographic study of schizophrenia. In: Mitsuda H et al (eds) Clinical genetics in psychiatry: problems in nowological classification. Igaku Shoin, Tokyo, pp 209–217
Averback P (1981) Structural lesions of the brain in young schizophrenics. Can J Neurol Sci 8:73–76
Bäumer H (1952) Untersuchungen am Nucleus medialis und lateralis thalami bei Schizophrenie. In: Proceedings of the first international congress of neuropathology, vol 3. Rosenberg and Sellier, Turin, pp 636–647
Bird ED, Spokes EGS, Iversen LL (1979) Increased dopamine concentration in limbic areas of brain from patients dying with schizophrenia. Brain 102:347–360
Bliss EL (1976) Neurologic features of schizophrenic processes. In: Wolfe S, Berle BB (eds) The biology of the schizophrenic process. Plenum, New York, pp 40–45
Bogerts B, Herzer M, Häntsch H (1982) The dopamine-containing cell groups of the mesencephalon: a quantitative evaluation in normals, schizophrenia, and Parkinson's disease. Neuroscience [Suppl] 7:30
Bogerts B, Häntsch J, Herzer M (1983) A morphometric study of the dopamine-containing cell groups in the mesencephalon of normals, Parkinson patients, and schizophrenics. Biol Psychiat 18:951–969
Boronow JJ, Pickar D, Van Kammen DP, Paul SM (1983) Normal ventricles and sulci in schizophrenic patients. Psychopharmacol Bull 19:581–586
Bowen DM, Smith CB, White P, Davison AN (1976) Neurotransmitter-related enzymes and indices of hypoxia in senile dementia and other abiotrophies. Brain 99:459–496
Bowen DM, Smith CB, White P, Goodhardt MJ, Spillane JA et al (1977) Chemical pathology of the organic dementias. I. Validity of biochemical measurements on human post-mortem brain specimens. Brain 100:397–426
Broser K (1949) Hirngewicht und Hirnprozess bei Schizophrenie. Arch Psychiat Nervenkr 182:439–449
Bruetsch WL (1952) Specific structural neuropathology of the central nervous system (rheumatic, demyelinating, vasofunctional, etc) in schizophrenia. Proceedings of the first international congress of neuropathology, vol 1. Rosenberg and Sellier, Torino pp 487–499
Buscaino VM (1920/21) La cause anatomo-patologiche della manifestione schizofreniche nella demenza precoce. Riv Patol Nerv Ment 25:197–225
Buchsbaum MS (1983) PET in schizophrenia and affective illness. Abstract of the 7th world congress of psychiatry, Vienna, 11–17 July 1983, p 527

Buttlar-Brentano K (1952) Pathohistologische Feststellungen am Basalkern Schizophrener. J Nerv Ment Dis 116:646–653

Candy JM, Perry RH, Perry EK, Irving D, Blessed G et al (1983) Pathological changes in the nucleus of Meynert in Alzheimer's and Parkinson's disease. J Neurol Sci 54:277–289

Christensen E, Møller JE, Faurbye A (1970) Neuropathological investigation of 28 brains from patients with dyskinesia. Acta Psychiatr Scand 46:14–23

Colon EJ (1972) Quantitative cytoarchitectonics of the human cerebral cortex in schizophrenic dementia. Acta Neuropathol (Berl) 20:1–10

Corsellis JAN (1976) Psychoses of obscure pathology. In: Blackwood W, Corsellis, J (eds) Greenfield's neuropathology, 3rd edn. Arnold, London, p 903

Crow JT (1982a) Two syndromes in schizophrenia. TINS 5:351–354

Crow JT (1982b) Neurohumoral and structural changes in schizophrenia. Prog Brain Res 55:407–418

Crow JT (1983) Schizophrenia as an infection. Lancet 1:819

Crow JT, Cross AJ, Johnstone EC, Owen F et al (1982) Changes in D_2 dopamine receptor numbers in post-mortem brain in schizophrenia in relation to the presence of type I syndrome and movement disorder. In: Collu R et al (eds) Brain peptides and hormones. Raven, New York, pp 43–53

Davison K, Bagley CR (1969) Schizophrenic-like psychoses associated with organic disorders of the central nervous system. In: Herrington RN (ed) Current problems in neuropsychiatry. Headley, Ashford, pp 113–184

Dewan MJ, Pandurangi AK, Lee SH, Ramachandran T, Levy B, Boucher M, Yozawitz A, Major LF (1983) Central brain morphology in chronic schizophrenic patients: a controlled study. Biol Psychiat 18:1133–1140

Dide MM (1934) Les syndromes hypothalamiques et la dyspsychogenèse. Rev Neurol (Paris) 6:941–943

Dom R, DeSaedeleer J (1981) Quantitative cytometric analysis of basal ganglia in catatonic schizophrenia. Abstract of the 3rd world congress of biological psychiatry, Stockholm, p 76

Donnely EF, Weinberger DR, Waldman IN et al (1980) Cognitive impairment associated with morphological brain abnormalities on CT in Chronic schizophrenic patients. J Nerv Ment Dis 168:305–308

Doyle JT, Price D, De Long MR (1983) Alzheimer's disease: a disorder of cortical cholinergic innervation. Science 219:1184–1190

Elvidge AR, Reed GE (1938) Biopsy studies of cerebral pathological changes in schizophrenia and manic-depressive psychosis. Arch Neurol Psychiatr 40:227–268

Farley IJ, Price HS, McCullough E et al (1978) Norepinephrine in chronic paranoid schizophrenia: above-normal levels in limbic forebrain. Science 200:456–458

Feighner JP, Robins E, Guze S, Woodruff RA, Winokur G, Munoz R (1972) Diagnostic criteria for use in psychiatric research. Arch Gen Psychiatry 26:57–62

Feuerlein W, Dilling H (1967) Echo-encephalographische Befunde bei Schizophrenie. Arch Psychiatr Nervenkr 209:137–146

Fishman M (1975) The brain-stem in psychosis. Br J Psychiatry 126:414–422

Fünfgeld E (1937) Bemerkungen zur Histopathologie der Schizophrenie. Z Ges Neurol Psychiatry 158:232–234

Glezer II, Soukhoroukova LI (1966) Les particularités structurelles de la neurologia dans la schizophrénia à évolution periodique et continue. Zh Nevropatol Psikhiatr 66:1529–1537

Gluck E, Radu EW, Hundt C et al (1980) A computed tomographie prospective study of chronic schizophrenics. Neuroradiology 20:167–171

Golden CJ, Moses JA, Zelazowski R, Graber B et al (1980) Cerebral ventricular size and neuropsychological impairment in young chronic schizophrenics. Arch Gen Psychiatry 37:619–623

Golden CJ, Graber B, Coffman J, Berg RA, Newlin DB, Bloch S (1981) Structural brain deficits in schizophrenia. Identification by CT scan density measurements. Arch Gen Psychiatry 38:1014–1017

Gross G, Huber G, Schüttler RL (1982) Computerized tomography studies on schizophrenic diseases. Arch Psychiatry Nervenkr 231:519–526

Hankhoff LD, Peress NS (1981) Neuropathology of the brain stem in psychiatric disorders. Biol Psychiatry 16:945–952

Hassler R (1967) Funktionelle Neuroanatomie und Psychiatrie. In Gruhle HW et al (eds) Psychiatrie der Gegenwart, vol 1. Springer, Berlin Heidelberg New York (1979)

Haug JO (1962) Pneumencephalographic studies in mental disease. Acta Psychiat Scand 38 (Suppl) 165:1–114

Heath RG, Franklin DE, Shraberg D (1979) Gross pathology of the cerebellum in patients diagnosed and treated as functional psychiatric disorders. J Nerv Dis 167:585–592

Heath RG, Franklin DE, Walker CF, Keating JW (1982) Cerebellar vermal atrophy in psychiatric patients. Biol Psychiatry 17:569–583

Hechst B (1931) Zur Histologie der Schizophrenie mit besonderer Berücksichtigung der Ausbreitung des Prozesses. Z Ges Neurol Psychiatry 134:163–267

Hempel KJ, Treff WM (1959) Über normale „Lücken" und „pathologische Lückenbildung" in einem subkortikalen Prisma (mediodorsaler Thalamuskern). Beitr Pathol Anat 121:288–292

Heyck H (1954) Kritischer Beitrag zur Frage anatomischer Veränderungen im Thalamus bei Schizophrenie. M Psychiatr Neurol 128:106–128

Hopf A (1952) Über histopathologische Veränderungen im Pallidum und Striatum bei Schizophrenie. In: Proceedings of the first international congress of neuropathology, vol 3. Rosenberg and Sellier, Turin, pp 629–635

Huber G (1957) Pneumencephalographische und psychopathologische Bilder bei endogenen Psychosen. Springer, Berlin Göttingen Heidelberg

Huber G (1961) Chronische Schizophrenie. Synopsis klinischer und neuro-radiologischer Untersuchungen an defektschizophrenen Anstaltspatienten. Hüthig, Heidelberg

Hyden H (1952) Nerve cell chemistry and neuropathological problems studied by means of quantitative methods. In: Proceedings of the first international congress of neuropathology, vol 3. Rosenberg and Sellier, Turin, pp 570–594

Ingvar DH (1980) Abnormal distribution of cerebral activity in chronic schizophrenia: a neurophysiological interpretation. In: Baxter C, Melnechuk T (eds) Perspectives in schizophrenia research. Raven, New York, pp 107–125

Ingvar DH, Franzen G (1974) Abnormalities of cerebral blood flow distribution in patients with chronic schizophrenia. Acta Psychiatr Scand 50:425–462

Jellinger K (1973) Axonal dystrophies. Prog Neuropathol 2:129–178

Jellinger K (1977) Neuropathologic findings after neuroleptic long-term therapy. In: Roizin L, Shiraki H, Grcevic N (eds) Neurotoxicology. Raven, New York, pp 25–42

Jellinger K (1980) Zur Neuropathologie schizophrener Psychosen. Curr Top Neuropathol 6:85–99

Jernigan TL, Zatz LM, Mojes JA, Berger PA (1982) Computed tomography in schizophrenics and normal volunteers. Arch Gen Psychiatry 39:765–770, 771–773

Johnstone EC, Crow TJ, Frith CD et al (1976) Cerebral ventricular size and cognitive impairment in chronic schizophrenia. Lancet 2:924–926

Josephy H (1923) Beiträge zur Histopathologie der Dementia praecox. Z Ges Neurol Psychiatr 86:391–485

Kingsley D, Trimble M (1978) Cerebral ventricular size in chronic schizophrenia. Lancet 1:278–279

Kirschbaum WR, Heilbrunn G (1944) Biopsies of brain of schizophrenic patients and experimental animals. Arch Neurol Psychiatry 51:155–162

Kraepelin E (1919) Dementia praecox and paraphrenia. Livingstone, Edinburgh

Lempke R (1935) Untersuchungen über die soziale Prognose der Schizophrenie unter besonderer Berücksichtigung des encephalographischen Befundes. Arch Psychiat Nervenkr 104:89–136

Luchins DJ, Wyatt DR (1983) Cerebral asymmetry and cerebellar atrophy in schizophrenia: a controlled post-mortem study. Am J Psychiatry 138:1501–1502

Luchins DJ, Weinberger DR, Wyatt R (1979) Schizophrenia: evidence of a subgroup with reversed cerebral asymmetry. Arch Gen Psychiatry 36:1309–1311

Malamud N (1967) Psychiatric disorder with intracranial tumours of limbic system. Arch Neurol 17:113–123

Malamud N (1975) Organic brain disease mistaken for psychiatric disorders. A clinico-pathologic study. In: Benson DF, Blumer D (eds) Psychiatric aspects of neurologic disease. Gruner and Stratton, New York, pp 287–310

Malamud N, Boyd D (1939) "Sudden" brain death in schizophrenia with extensive lesions in the cerebral cortex. Arch Neurol Psychiatry 41:352–364

Meyer A (1952) Critical evaluation of histopathological findings in schizophrenia. In: Proceedings of the first congress of neuropathology, vol 1. Rosenberg and Sellier, Turin, pp 649–666

Miskolczy D (1937) Die örtliche Verteilung der Gehirnveränderungen bei Schizophrenie. Z ges Neurol Psychiatr 158:203–208

Miyakawa T (1964) A histopathological study on the brains of schizophrenia. Findings in the limbic system. Folia Psychiatr Neurol (Jpn) 66:937–942

Miyakawa T, Sumiyoshi S, Deshimaru M et al (1972) Electron microscopic study on schizophrenia: mechanisms of pathological changes. Acta Neuropathol (Berl) 20:67–77

Mundt CH, Radü W, Gluck E (1980) Computertomographische Untersuchungen der Liquorräume an chronisch schizophrenen Patienten. Nervenarzt 51:743–748

Münzer FT (1925) Beiträge zur Pathologie und Pathogenese der Dementia praecox (Schizophrenie). Z Neurol Psychiatr 103:73–132

Naeser MA, Levine ML, Benson DF et al (1981) Frontal leukotomy size and hemispheric asymmetries on CT scans of schizophrenics with variable recovery. Arch Neurol 38:36–37

Nagasaka G (1925) Zur Pathologie der extrapyramidalen Zentren bei Schizophrenie. Arb Neurol Inst Univ Wien 27:363–396

Nagy K (1963) Pneumenzephalographische Befunde bei endogenen Psychosen. Nervenarzt 34:543–548

Nasrallah HA, Jacob CG, Calley-Whitters MM, Kuperman S (1982) Cerebral ventricular enlargement in subtypes of chronic schizophrenia. Arch Gen Psychiatr 39:774–777

Nasrallah HA, Jacob CG, Kuperman S et al (1983) A histological study of the corpus callosum in chronic schizophrenia. Psychiatry Res 8:251–260

Newlin DB, Carpenter B, Golden CJ (1981) Hemispheric asymmetries in schizophrenia. Biol Psychiatry 16:561–581

Nieto D, Escobar A (1972) Major psychoses. In: Minckler J (ed) Pathology of the nervous system, vol 3. McGraw-Hill, New York, pp 2654–2665

Nyback H, Berggren BM, Hindmarsh T (1982) Computed tomography of the brain in patients with acute psychoses and in healthy volunteers. Acta Psychiat Scand 65:29–34

Owens DGC, Johnstone EC, Bydder GM, Kreel L (1980) Unsuspected organic disease in chronic schizophrenia demonstrated by computed tomography. J Neurol Neurosurg Psychiatr 12:1065–1069

Palma EC, Sotelo JR (1952) Anatomo-clinical and histological investigation on schizophrenia. Proceedings of the first international congress of neuropathology, vol 1. Sellier and Rosenberg, Torino, pp 637–647

Pearlson GD, Veroff AE, Mc Hugh PR (1981) The use of computed tomography in psychiatry. John Hopkins Med J 149:194–202

Penn H, Racy J, Lapham L, Mandel M, Sandt J (1972) Catatonic behavior, viral encephalopathy, and death. Arch Gen Psychiatry 27:758–761

Perry EK, Perry RH (1983) Human brain neurochemistry – some postmortem problems. Life Sci 33:1733–1743

Peters G (1937) Zur Frage der pathologischen Anatomie der Schizophrenie. Z Ges Neurol Psychiatry 160:361–380

Peters G (1956) Dementia praecox und manisch-depressives Irresein. In: Uehlinger E (ed) Handbuch der speziellen pathologischen Anatomie und Histologie, 13/4.

Peters G (1967) Neuropathologie und Psychiatrie. In: Kinsker KP, Meyer SE, Müller C, Strömgren E (eds) Psychiatrie der Gegenwart, vol 1. Springer, Berlin Heidelberg New York, pp 286–324

Peters G (1974) Pathologische Anatomie der endogenen Psychosen. In: Ergebnisse biologischer Forschung bei endogenen Psychosen. Das ärztliche Gespräch, vol 23. Tropon, Köln, pp 11–21

Riederer P, Jellinger K (1982) Biochemie und morphologische Aspekte der Schizophrenie. Schwerpunktmedizin 5:32–40

Riederer P, Jellinger K (1983) Morphologie und Pathobiochemie der Parkinson-Krankheit. In: Gänshirt M (ed) Pathophysiologie, Klinik und Therapie des Parkinsonismus, Ed Roche, Basel pp 31–50

Rieder RO, Donnelly EF, Herdt JR et al (1979) Sulcal prominence in young chronic schizophrenic patients. CT scan findings associated with impairment on neuropsychological tests. Psychiatry Res 1:1–8

Rosenthal R, Bigelow LB (1972) Quantitative brain measurements in chronic schizophrenia. Br J Psychiatry 121:259–264

Rowland LP, Mettler FA (1949) Cell concentration and laminar thickness in the frontal cortex of psychotic patients. J Comp Neurol 90:255–265

Scharenberg K, Brown EO (1954) Histopathology of catatonic states. A study with silver carbonate. J Neuropathol. Exp Neurol 13:592–600

Scheibel AB, Kovelman JA (1980) Disorientation of the hippocampal pyramidal cells and its processes in schizophrenic patients. Biol Psychiat 16:101–102

Scheller H (1966) Morphologische Aspekte bei schizoformen Psychosen. Arzneimittelforsch 16:279–281

Shulack NR (1945) Exhaustion syndrome in excited psychotic patients. Am J Psychiatry 102:466–475

Spielmeyer W (1930) Die anatomische Krankheitsforschung in der Psychiatrie. In: Bumbke O (ed) Handbuch der Psychiatrie, vol 10/7. Springer, Berlin, p 1

Spielmeyer W (1931) The problem of the anatomy of schizophrenia. Res Publ Assoc Res Nerv Ment Dis 10:105–110

Spokes EGS (1979) An analysis of factors influencing measurements of dopamine, noradrenaline, glutamate decarboxylase and choline acetylase in human post-mortem brain tissue. Brain 102:333–346

Stauder E (1934) Lethale Catatonia. Arch Psychiatr Nervenkr 102:614–634

Stevens JR (1973) An anatomy of schizophrenia? Arch Gen Psychiatry 20:177–189

Stevens JR (1982) Neuropathology of schizophrenia. Arch Psychiatry 39:1131–1139

Stevens JR, Albrecht P, Godfrey L, Krauhammer E (1983) Viral antigen in the brain of schizophrenic patients? A preliminary report. Adv Biol Psychiatr 12:76–96

Storey PB (1966) Lumbar i ar encephalography in chronic schizophrenia: a controlled experiment. Br J Psychiatr 112:135–144

Takahashi R, Inaba K, Kato N et al (1982) CT scanning and the investigation of schizophrenia. In: Jansson B, Perris C, Struwe G (eds) Biological psychiatry 1981. Elsevier-North Holland, New York, pp 259–268

Tanaka Y, Hazama H, Kawahara R, Kobayashi K (1981) Computerized tomography of the brain in schizophrenic patients. Acta Psychiatr Scand 63:191–197

Tatetsu S (1964) A contribution to the morphological background of schizophrenia. With special reference to findings in the telencephalon. Acta Neuropathol (Berl) 3:558–571

Tatetsu S (1974) On histologic findings in schizophrenia and schizophrenic state. In: Mitsuda H, Fukuda T (eds) Biological mechanisms of schizophrenia and schizophrenia-like psychoses. Igaku-Shoin, Tokyo, pp 288–289

Van der Horst L (1952) Histopathology of clinically diagnosed schizophrenic psychoses or schizophrenia-like psychoses of unknown origin. In: Proceedings of the international congress of neuropathology, vol 3. Rosenberg and Sellier, Turin, pp 648–659

Vogt C, Vogt O (1952) Alterations anatomiques de la schizophrenia et d'autres psychoses dites fonctionelles. In: Proceedings of the first international congress of neuropathology vol 1. Rosenberg and Sellier, Torino, pp 515–532

Wahren NL (1952) The changes of hypothalamic nuclei in schizophrenia. In: Proceedings of the first International congress of neuropathology, vol I. Rosenberg and Sellier, Torino, p 660

Weinberger DR, Wyatt RJ (1980) Structural brain abnormalities in chronic schizophrenia: computed tomography findings. In: Baxter C, Melnechuk T (eds) Perspectives in schizophrenia research. Raven, New York, pp 29–38

Weinberger DR, Torrey EF, Neophytides AN et al (1979) Lateral cerebral ventricular enlargement in chronic schizophrenia. Arch Gen Psychiatry 36:735–739

Weinberger DR, Bigelow LB, Kleinman JE et al (1980) Cerebral ventricular enlargement in chronic schizophrenia. Arch Gen Psychiatr 37:11–14

Weinberger DR, DeLisi LE, Perman GP et al (1982) Computed tomography in schizophreniform disorder and other acute psychiatric disorders. Arch Gen Psychiatry 39:778–783

Weinberger DR, Wagner RL, Wyatt RJ (1983) Neuropathological studies of schizophrenia: A selective review. Schizophrenia Bull 9:193–212

Whitehouse PJ, Hendreen JC, White CL, Price DL (1983) Basal forebrain neurons in the dementia of Parkinson's disease. Ann Neurol 13:243–248

Wildi E, Linder A, Costooles G (1967) Schizophrenie et involution cérébrale sénile. Psychiatr Neurol (Basel) 154:1–26

The Significance of Psychopathological Classification in Interpreting Biochemical Findings

E. GABRIEL

Introduction

The empirical background of this paper is a research program in schizophrenia involving neuropathology, biochemistry, and clinical psychiatry (Reynolds et al. 1980, 1981 a, b, c; Riederer et al., to be published). It is a postmortem study of neurochemical abnormalities in brain tissue of patients who in most cases died after a long stay in a psychiatric hospital. The three disciplines represented and the special design of a postmortem study should be stressed because these facts lead to problems of case definition and identification and identify certain preconditions of the study which may perhaps influence its results. I should like to mention only two points dependent on the design of this study which are related closely to the topic of my communication:

1. The probands were, in general, very longstay mental hospital patients and thus had not only been chronically ill but also belonged to a risk population for institutionalism. From a functional viewpoint and with respect to the cross-sectional evaluation of the patients' behavior and symptoms, it may be difficult to distinguish between illness-dependent and situation-dependent phenomena. A classic example of this problem is the so-called residual state or defect in schizophrenics in which paranoid symptoms must also be taken into account. The solution of the problem may be that in the case of functional psychoses, a psychopathological definition can only be a functional one which does not per se represent etiological conditions of the described state. (In contrast, the notion of organic psychoses comprises an etiological inclusion criterion.)

2. Another precondition of this type of study concerns the stability of the psychopathological states of the patients before death. (Of all probands, 80% have been stable in the last 3 months of life.) It is an important finding in psychophysiology, and is perhaps also true for this field, that physiological changes emerge mainly when psychological ones take place but that in stable psychological states, even when these are abnormal, physiological parameters remain normal.

Thus the study relates to biochemical characteristics of probands who had been labeled schizophrenics because of the more or less typical symptoms, which refer to psychological functioning. However, one has to bear in mind that this depends on various factors, including situational and treatment factors, which are probably not independent of each other.

Classification

The major parameter under investigation was spiroperidol binding to dopamine receptors in brain tissue of patients in whom schizophrenia had been diagnosed and who died in a mental hospital. Any neuropathological changes other than senile ones were excluded. In defining schizophrenia we started from the basis of the ICD diagnosis but, using a polydiagnostic approach, compared it to four other systems of diagnosis for schizophrenic psychoses: the criteria of Bleuler, those of Schneider, the Vienna Research Criteria of Berner, and Feighner's criteria (WPA Diagnostic Criteria for Schizophrenic and Affective Psychoses 1983). For details, see Table 1.

The comparison shows:

1. That the diagnoses using Bleuler's criteria were practically the same as those using ICD. In fact, the ICD concept of diagnosing schizophrenia is Bleuler's concept.

2. There is a large overlap with the diagnoses using Schneider's first-rank symptoms. With these criteria a few of the ICD and Bleuler schizophrenics are not included. This is not surprising. Schneider himself did not base his clinical diagnosis of schizophrenia exclusively on first-rank symptoms. Nevertheless, in our material 80% of the ICD schizophrenics show first-rank symptoms at some time during the course of their illness.

3. The Vienna Research Criteria of Berner are psychopathological criteria, as are the ICD, Bleuler, and Schneider criteria. We used these Vienna criteria without regarding affective blunting as a criterion, thus basing the classification only on formal thought disorders and neologisms. (In this study affective blunting is not regarded as a diagnostic criterion because of the great difficulty in its evaluation, particularly in long-term inpatients with the affective impoverishment resulting from life in a large mental hospital.) Using this variant of the Vienna criteria about one-half (48%) of the ICD schizophrenics are regarded as positive, the criteria again occurring at any time during the illness.

4. The Feighner criteria are not only psychopathological criteria; in them a time criterion (chronicity) is also taken into consideration. It is not surprising that these criteria are positive in the majority (74%) of long-stay inpatients.

Table 1. ICD 295/297 and four other diagnostic systems for schizophrenic psychoses

ICD 295/297	Bleuler	Schneider	Berner	Feighner
31	30	25	15	23
100%	96%	80%	48%	74%
Feighner-pos.				
23	23	21	15	
100%	100%	91%	65%	
Feighner-neg.				
8	7	4	0	
100%	87%	50%		

Otherwise in our study the actual psychopathological state is only considered by the ICD fourth digit. For all other diagnostic systems the total observation time has been taken into consideration.

If one uses the time criterion in Feighner's system in addition to ICD, differences in the overlap with the other systems emerge, namely an increase in overlap with the groups of Schneider- and Berner-positive cases and a decreased overlap between Feighner-negative ICD schizophrenics and Schneider- and Berner-positive schizophrenics. (For details, see Table 1.) Statistically, this may indicate a relationship between these sets of psychopathological criteria and chronicity. But of course this is indicative of the fact that the criteria are not always present throughout the illness, and particularly that the Berner criteria occur more frequently in the later stages of the disease than at the beginning. Thus the change of meeting certain symptomatological diagnostic criteria grows with the observation time (Gabriel 1978).

Therefore, in our design these psychopathological differentiations seemed to be meaningful:

1. A syndromatological one related to the present state, including a large group of patients (ICD 295.3/297 versus 295.6)
2. A psychopathological one which is much more restrictive but neglects the present state (Berner-positive versus Berner-negative)
3. A differentiation including a time criterion (Feighner-positive versus Feighner-negative).

Results

Here no emphasis is made upon the differences between patients treated with neuroleptics within 3 months before death and patients who had no neuroleptic treatment in this period. Nor is mention made of the problem whether the degree of neuroleptic drug treatment in terms of dosage and/or time is influential (Reynolds et al. 1981). This is a problem which, for methodological reasons, not easily solved (problem of equivalence of doses of different neuroleptics, problem of different distribution of the medication in time).

Finally, it is not the aim of this paper to present in detail data which are published elsewere (Riederer et al., to be published). It is intended that we focus on the significance of psychopathological classification in interpreting the biochemical findings, using this study only as an example. Table 2 lists the mean values of B_{max} and K_D and, instead of the standard deviation, the raw minima and maxima in order to point out the great variations, particularly in the groups on neuroleptic treatment.

The low receptor number in all those groups without neuroleptic drug treatment within the last 3 months before death is particularly notable. In any case, it relates to patients who were in a stable state at that time, only few of them showing paranoid symptoms. (It is well known to psychiatrists that paranoid symptoms in chronically ill patients may not be based on the same pathogenetic pathways as those in acute psychoses.) The Berner-positive probands, too, are few. At the time of investigation they all showed a residual state (ICD 295.6), as did the majority of the Berner-nega-

Table 2. ^{3}H-Spiroperidol binding in postmortem putamen, neuroleptic drug treatment, and psychopathological classification

	n	B_{max} (pmol/g)	K_D (nM)
Controls	22	21.9±2.0	0.15±0.03
ICD 295.3/297			
On NL	13	26.3 (*11.7–41.5*)	0.45 (*0.17–1.50*)
Without NL	2	13.8 (13.2–14.4)	0.13 (0.09–0.17)
ICD 295.6			
On NL	7	22.8 (*11.7–50.7*)	0.93 (*0.09–3.10*)
Without NL	7	16.4 (8.7–23.9)	0.14 (*0.07–0.27*)
Berner-pos.			
On NL	11	28.3 (*12.3–50.7*)	0.84 (*0.10–3.10*)
Without NL	3	20.1 (18.3–21.4)	0.13 (0.11–0.15)
Berner-neg.			
On NL	9	21.2 (*11.7–41.5*)	0.45 (*0.09–1.50*)
Without NL	7	14.0 (8.7–23.9)	0.21 (*0.07–0.65*)
Feighner-pos.			
On NL	17	25.2 (*11.7–50.7*)	0.66 (*0.09–3.10*)
Without NL	7	16.4 (8.7–23.9)	0.11 (0.07–0.17)
Feighner-neg.			
On NL	3	24.6 (*14.1–41.5*)	0.67 (*0.17–1.50*)
Without NL	3	14.6 (13.2–16.2)	0.34 (*0.09–0.65*)

Italics indicates variation exceeding that in controls

tive cases (except one with senile dementia and another with a paranoid syndrome) and the majority of the Feighner-positive probands (except one with senile dementia).

Thus from a psychopathological viewpoint it seems to be possible that the different distribution of the biochemical measures is based (a) on treatment effects and (b) on pairs of psychopathological characteristics such as productive versus non-productive/defective, acute versus chronic, unstable versus stable, characterizing morbid states which are probably interrelated with neuroleptic drug treatment. (From this point of view the next steps of our research in this field have to focus on quantitative aspects of this kind and to apply multivariate statistical methods.)

Conclusion

The research situation in this field reminds one of a similar one some 30 years ago. Summarizing the important attempts to find out the biological foundation of schizophrenia by means of the (then novel) techniques of clinical endocrinology, Bleuler (1954) admitted that this purpose had not been achieved. But he stressed that, instead of an endocrinology of schizophrenia, important traits of a physiology of emotionality had been found. It would appear that the same is true at the synaptic level.

References

Bleuler M (1954) Endokrinologische Psychiatrie. Thieme, Stuttgart

Gabriel E (1978) Die langfristige Entwicklung von Spätschizophrenien. Karger, Basel

ICD (1978) Mental disorders. Glossary and guide to their classification in accordance with the 9th revision of the international classification of diseases. WHO, Geneva

Reynolds GP, Reynolds LM, Riederer P, Jellinger K, Gabriel E (1980) Dopamine receptors and schizophrenia, drug effect or illness. Lancet II: 1251

Reynolds GP, Riederer P, Gabriel E (1981a) Propranolol binding in human brain. In: Riederer P, Usdin E (eds) Transmitter biochemistry of human brain tissue. Macmillan, London, pp 105–112

Reynolds GP, Riederer P, Jellinger K, Gabriel E (1981b) Dopamine receptors and schizophrenia: the influence of neuroleptic drug treatment and disease syndromes. In: Perris C, Struwe G, Jansson B (eds) Biological psychiatry. Elsevier, Amsterdam, pp 715–718

Reynolds GP, Riederer P, Jellinger K, Gabriel E (1981c) Dopamine receptors and schizophrenia: the neuroleptic drug problem. Neuropharmacology 20: 1319–1320

Riederer P, Jellinger K, Gabriel E (to be published) ^{3}H-Spiperone binding to post mortem human putamen in paranoid and nonparanoid schizophrenics. In: Berner P (ed) Proceedings of the VII world congress of psychiatry. Plenum, London

WpA diagnostic criteria for schizophrenic and affective psychoses (1983) American Psychiatric Press, Washington

Receptors, Neuroleptics and Dopamine Concentrations in Schizophrenia – Postmortem Studies of Human Brain Tissue

G. P. REYNOLDS

Introduction

There is little doubt that biochemical investigation of human postmortem brain tissue has made a substantial contribution to the understanding and treatment of disease. It was as a direct result of the observation of a dopamine deficit in the corpus striatum of Parkinson's disease patients that L-dopa was introduced and found to be so successful (Birkmayer and Hornykiewicz 1962). Unfortunately, such a valuable advance has yet to be made in other fields of neuropsychiatry, although recently some potentially important findings have been made towards our understanding of schizophrenia and the mechanisms of antipsychotic drugs.

Increased Dopamine Receptors in Schizophrenia

The most rigorously investigated hypothesis of the neurochemistry of schizophrenia is that the disease derives from a hyperactivity of dopamine neurotransmission. Concentrations of the transmitter, its metabolites, and associated enzymes have all been investigated using postmortem brain tissue from schizophrenics, albeit with few conclusive and consistent results. Recently, work has focused on a more consistently reported finding: an increase in the number of receptors to dopamine in postmortem brain tissue. Initial work from Crow and his colleagues (Owen et al. 1978) suggested that the increase is a function of the disease process, since they report an increase which appears independent of antischizophrenic neuroleptic medication. Other groups, however, find an increase only in drug-treated patients (Mackay et al. 1980, 1982; Reynolds et al. 1981). Certainly, animal experiments also show an in-

Table 1. ^{3}H-Spiperone binding in human putamen

	$B_{max}(pmol/g^{-1})$
Controls (13)	22.7 ± 1.2
Schizophrenics	
Chronic neuroleptic treatment (8)	32.2 ± 4.5*
Short-term or no neuroleptic treatment (7)	17.7 ± 1.4

* $p < 0.05$ vs controls

crease in these receptors due to long-term neuroleptic medication (e.g. Clow et al. 1980).

In previous work we have attempted to answer the question of whether these dopamine D_2 receptors are increased due to the disease process or to the drug treatment of schizophrenics. Our findings are summarized in Table 1. B_{max} values (which correspond to receptor number and were obtained by Scatchard analysis of ligand binding to membrane preparations) are significantly increased in patients who received long-term (generally several years) neuroleptic treatment before death. On the other hand, we were able to identify a group of patients, all with less than 3 months neuroleptic treatment in the last year of life, who had no abnormal increase in receptor number. In fact, the three patients receiving no such treatment had a significantly lower number of D_2 receptors (15.4 ± 1.7 pmol/g^{-1} tissue). The increase in long-term-treated patients was found to be independent of any disease subclassification, i.e. whether they were paranoid (ICD 295.3) or residual, defect state schizophrenics (ICD 295.6).

Thus the effect would appear to be independent of the disease process and purely an effect of long-term neuroleptic medication. Certainly, there is other evidence that neuroleptics increase D_2 receptor number in man. Samples studied from two neuroleptic-treated patients exhibiting the neuropathological changes of Alzheimer's disease also had a substantial and significant increase in spiperone binding sites (Reynolds et al. 1982 b).

Site Specificity of Neuroleptic Action

The observation that extrapyramidal side-effects do not occur to the same extent with all neuroleptic drugs is generally thought to be due to inherent anticholinergic activity, which is much greater in some neuroleptics (e.g. thioridazine) than in others (Snyder et al. 1974). However, the theory that the mesolimbic dopamine pathway functions abnormally in schizophrenia (Stevens 1979) has led some workers (Borison et al. 1981) to propose a "site-specific blockade of dopamine receptors" and to claim that thioridazine is several orders of magnitude more active on limbic than on striatal dopamine receptors. We have investigated this claim using postmortem tissue from the accumbens and putamen of human brain. The results, shown in Table 2, indicate no substantial differences between the inhibition of

Table 2. Potencies of neuroleptics as inhibitors of specific ^{3}H-spiperone binding in striatum (putamen) and limbic system (nucleus accumbens)

	IC_{50} (nmol/l)	
	Putamen	N. accumbens
Thioridazine	170	200
Haloperidol	46	36
Chlorpromazine	140	140

[3]H-spiperone binding to receptors in the two tissues (Reynolds et al. 1982a). Thus the evidence suggests that it is incorrect to invoke a "site-specific" action for thioridazine on dopamine receptors in limbic areas of human brain.

5-Hydroxytryptamine$_2$ Receptors in Schizophrenia

The lack of any definitive indication of dopamine dysfunction in schizophrenia has provoked a return of attention to the other monoamine neurotransmitters, noradrenaline (Hornykiewicz 1982) and 5-hydroxytryptamine (5-HT) (Trulson and Jacobs 1979). Originally, dopamine was especially implicated in schizophrenia, as this neurotransmitter was clearly involved in the behavioural effects of amphetamine, which, in high or chronic dosage, is able to induce a psychosis in man reminiscent of acute paranoid schizophrenia. However, both noradrenaline and 5-HT are involved in some components of amphetamine-induced behaviour. In particular, large doses or chronic administration of amphetamine to animals induces effects which can be blocked by 5-HT antagonists; this is referred to as the 5-HT behavioural syndrome (Peroutka et al. 1981). It is mediated by 5-HT$_2$ receptors, to which many neuroleptic drugs and hallucinogenic compounds, such as LSD, bind with high affinity.

There have been reports of changes in 5-HT$_2$ receptors in schizophrenia. Bennett et al. (1979) found an apparent decrease in ^{3}H-LSD binding to frontal cortex in schizophrenia. However, Whitaker et al. (1981) were unable to confirm this and in fact reported a significant increase in maximal ^{3}H-LSD binding in schizophrenics who were thought to have been free of neuroleptic drugs.

In an attempt to shed more light on the possible involvement of 5-HT$_2$ receptors in schizophrenia we have studied them using a newly developed ligand, ^{3}H-ketanserin, which binds with high affinity and greater specificity than LSD to this receptor population (Reynolds et al. 1983). The results (Table 3) indicate no difference either in receptor number (which should approximate to binding at 2 nM) or receptor affinity (a function of the binding ratio). Thus the results are certainly not consistent with those of Bennett et al. (1979).

Further evidence has been obtained which argues against an important role for these receptors in the action of antipsychotic drugs. While many neuroleptics are found to have a high affinity for the 5-HT$_2$ receptor, Table 4 indicates that this affinity in no way parallels that for the D$_2$ site in human brain tissue (Reynolds 1983a) and thus bears little relationship to antipsychotic potency. What is notable is that some drugs (e.g. chlorpromazine) have a relatively higher affinity for the 5-HT$_2$ than

Table 3. ^{3}H-Ketanserin binding to human cortex preparations

	Binding at 2 nM[a]	Binding ratio[b]
Controls	6.73±0.53	0.531±0.024 (n = 10)
Schizophrenics	8.00±0.70	0.485±0.023 (n = 11)

[a] Mean ± SE in pmol/g tissue
[b] Ratio of binding with 0.4 nM and 2 nM ketanserin

Table 4. Neuroleptic drug displacement of ligand binding to 5-hydroxytryptamine$_2$ (ketanserin) and dopamine D$_2$ (spiperone) receptors

	$K_I(nM)$	
	^{3}H-Ketanserin (cortex)	^{3}H-Spiperone (striatum)
Fluphenazine	4	1
Haloperidol	31	5.8
Chlorpromazine	2	18
Thioridazine	26	21
Clozapine	3.7	105

the D$_2$ receptor. Thus doses of chlorpromazine which will block transmission via the dopamine receptor labelled by spiperone (which is presumably the locus of its antipsychotic effect) may also be expected to saturate cortical 5-HT$_2$ receptors. More data are required before a relationship with any side-effect of phenothiazine treatment can be established, although the occurrence of a physiological correlate of such a strong neurochemical influence would seem likely.

Clearly, 5-HT has not yet replaced dopamine in the forefront of research into the neurochemical aetiology of schizophrenia. In fact, some very recent evidence with relatively large sample groups indicates that there are indeed changes in presynaptic dopamine function in schizophrenia as reflected by increased dopamine concentrations.

Limbic Dopamine in Schizophrenia

While some studies have been unable to identify a significant increase in dopamine concentrations in postmortem brain tissue from schizophrenics (Crow et al. 1981), in two reports of studies from a large series of cases (Bird et al. 1979; Mackay et al. 1982) dopamine has been found to be increased, particularly in limbic regions. Mackay et al. (1982) find this increase to be most marked in younger patients and not obviously related to neuroleptic medication.

As mentioned above, the limbic system is particularly implicated in schizophrenia (Stevens 1979). The temporal lobe is possibly involved, considering the psychotic behaviour often seen in temporal lobe epilepsy (Flor-Henry 1969). One limbic region of the temporal lobe which receives a dopaminergic innervation is the amygdala, which has recently been studied in postmortem brain tissue (Reynolds 1983 b): a substantial increase in dopamine concentrations has been found. Table 5 shows this to be regionally and biochemically specific, since caudate dopamine and amygdala noradrenaline are not significantly changed. Finally, it is apparent that the increase is restricted to the left temporal lobe, an observation which provides the first biochemical support for the wide range of psychological and neurophysiological evi-

Table 5. Catecholamine neurotransmitter concentration in bilaterally dissected postmortem brain tissue

	Left	Right
Amygdala		
Dopamine		
Schizophrenics	96.1*	55.7*
Controls	57.3	50.8
Noradrenaline		
Schizophrenics	58.7	61.5
Controls	54.2	63.2
Caudate nucleus		
Dopamine		
Schizophrenics	2185	2612
Controls	2572	2121

Values are geometric means in nanograms per gram tissue from 16 schizophrenic and 14 control brains.
* $p < 0.001$ by t-test of log-transformed data; other control-schizophrenic comparisons are not significant

dence for temporal lobe lateralization in schizophrenia (reviewed by Gruzelier 1981). Thus it would appear that further study of mesolimbic dopaminergic pathways, particularly those projecting to the amygdala, is indicated and may eventually serve to combine the various biochemical, neurophysiological and psychological interpretations of the dysfunction in schizophrenia into a single comprehensive hypothesis.

References

Bennett JP, Enna SJ, Bylund DB, Gillin JC, Wyatt RJ, Snyder SH (1979) Neurotransmitter receptors in frontal cortex of schizophrenics. Arch Gen Psychiatry 36:927–934

Bird ED, Spokes EGS, Iversen LL (1979) Increased dopamine concentration in limbic areas of brain from patients dying with schizophrenia. Brain 102:347–360

Birkmayer W, Hornykiewicz O (1962) Der L-3,4-Dioxyphenylalanin (L-Dopa)-Effekt beim Parkinson-Syndrom des Menschen. Arch Psychiatr Nervenkr 203:560–574

Borison RL, Fields JZ, Diamond RI (1981) Site specific blockade of dopamine receptors by neuroleptic agents in human brain. Neuropharmacology 20:1321–1322

Clow A, Theodorou A, Jenner P, Marsden CD (1980) Changes in rat striatal dopamine turnover and receptor activity during one year's neuroleptic administration. Eur J Pharmacol 63:135–144

Crow TJ, Owen F, Cross AJ, Ferrier N, Johnstone EC, McCreadie RM, Owens DGC, Poulter M (1981) Neurotransmitter enzymes and receptors in post mortem brain in schizophrenia: evidence that an increase in D_2 dopamine receptors is associated with the type I syndrome. In: Riederer P, Usdin E (eds) Transmitter biochemistry of human brain tissue. Macmillan, London, pp 85–96

Flor-Henry P (1969) Psychosis and temporal lobe epilepsy: a controlled investigation. Epilepsia 10:363–395

Gruzelier JH (1981) Cerebral laterality and psychopathology: fact and fiction. Psychological Med 11:219–227

Hornykiewicz O (1982) Brain catecholamines in schizophrenia – a good case for noradrenaline. Nature 299:484–486

Mackay AVP, Bird ED, Spokes EL, Rossor M, Iversen LL, Creese I, Snyder SH (1980) Dopamine receptors and schizophrenia: drug effect or illness? Lancet II:223–225

Mackay AVP, Iversen LL, Rossor M, Spokes E, Bird E, Arregui A, Creese I, Snyder SH (1982) Increased brain dopamine and dopamine receptors in schizophrenia. Arch Gen Psychiatry 39:991–997

Owen F, Crow TJ, Poulter M, Cross AJ, Longden A, Riley GJ (1978) Increased dopamine-receptor sensitivity in schizophrenia. Lancet II:915–916

Peroutka SJ, Lebowitz RM, Snyder SH (1981) Two distinct serotonin receptors with different physiological functions. Science 212:827–829

Reynolds GP (1983a) (^{3}H)-Ketanserin binding to 5-HT$_2$ receptors in human brain. Br J Pharmacol 78:273 p

Reynolds GP (1983b) Increased concentrations and lateral asymmetry of amygdala dopamine in schizophrenia. Nature 305:527–529

Reynolds GP, Riederer P, Jellinger K, Gabriel E (1981) Dopamine receptors and schizophrenia: the neuroleptic drug problem. Neuropharmacology 20:1319–1320

Reynolds GP, Cowey L, Rossor M, Iversen LL (1982a) Thioridazine is not specific for limbic dopamine receptors. Lancet II:499–500

Reynolds GP, Riederer P, Jellinger K, Gabriel E (1982b) Effects of neuroleptic treatment and disease state on dopamine receptors in post-mortem schizophrenic brain. Neuroscience 7:S177

Reynolds GP, Rossor MN, Iversen LL (1983) Preliminary studies of human cortical 5-HT$_2$ receptors and their involvement in schizophrenia and neuroleptic drug action. J Neural Trans [Suppl] 18:273–277

Snyder S, Greenberg D, Yamamura HI (1974) Antischizophrenic drugs and brain cholinergic receptors. Arch Gen Psychiatry 31:58–61

Stevens JR (1979) Schizophrenia and dopamine regulation in the mesolimbic system. Trends Neurosci 2:102–105

Trulson ME, Jacobs BL (1979) Long-term amphetamine treatment decreases brain serotonin metabolism: implications for theories of schizophrenia. Science 205:1295–1297

Whitaker PM, Crow TJ, Ferrier IN (1981) Tritiated LSD binding in frontal cortex in schizophrenia. Arch Gen Psychiatry 38:278–280

Brain Biochemistry in Schizophrenia: An Assessment

P. RIEDERER and G. P. REYNOLDS

In 1896 Emil Kraepelin distinguished "dementia praecox" from the affective psychoses. Shortly afterwards, Bleuler proposed the name "schizophrenia", and since that time, the beginning of modern biological psychiatry, studies have been performed with the aim of understanding the basic mechanisms of these somatic disturbances (Kraepelin 1919; Bleuler 1923). While genetic factors have indicated a hereditary contribution to the aetiology of schizophrenia (Tsuang 1976), the search for distinct pathophysiological associations with this disease has, until very recently, proved disappointing. Nevertheless, the results of some investigations, particularly in the fields of biochemistry and pharmacology, have provided us with a pointer to the changes at the level of brain chemistry which may underly this disease. This progress has been made with three overlapping approaches:

1. The model psychoses
2. The pharmacology of the antipsychotic drugs
3. Human postmortem brain studies.

Model Psychoses

Continued administration of amphetamine to man induces a pharmacotoxic psychosis the symptoms of which greatly resemble those of paranoid schizophrenia. As amphetamine releases catecholamines from storage sites and can block intraneuronal monoamine oxidase (MAO) and thus stimulates postsynaptic receptors, it has been suggested that these pharmacological properties may indicate the pathogenesis of psychotic symptoms. In line with such an assumption are clinical observations showing an enhancement of paranoid symptoms after administration of amphetamine to schizophrenic patients (Janowsky et al. 1973a). Furthermore, at high doses amphetamine releases serotonin as well as affecting the cholinergic system, as indicated by the beneficial effects of cholinergic drugs (Janowsky et al. 1973b).

However, amphetamine psychosis – as all other known models of schizophrenia – resembles only one facet of the heterogeneity of the natural disease (Carlsson 1978; Snyder 1972). This is even more true for psychedelic drugs like LSD, psilocybin and mescaline, as well as for those drugs which may induce pharmacotoxic psychosis in parkinsonian patients (L-dopa, amantadine, dopaminergic agonists, MAO inhibitors and anticholinergics). It might be expected that L-dopa would enhance the psychotomimetic effects of amphetamine; however, this seems not to be the case (Friedhoff and Alpert 1973). L-Dopa therapy can produce a wide range of psychotic symptoms and even depression in Parkinson's disease, as well as activating retarded depressions, turning bipolar depression into mania and effecting an increased agressivity in other depressed patients (Luchins 1975). Although the role of dopamine in these symptoms and side-effects might be more important, L-dopa can also reduce brain serotonin and interferes with gamma-aminobutyric acid (GABA) and acetylcholine.

Another example is cocaine, which blocks the reuptake of dopamine into the neuron but, like amphetamine, can also affect noradrenergic, serotonergic and cholinergic neuronal activity. Depending on dose, route, duration and clinical predisposition, cocaine can produce symptoms mimicking mania, schizophrenia or depression (Post et al. 1975).

Pharmacology of Antipsychotic Drugs

One of the first relevant approaches towards a "biology of schizophrenia" was the discovery that reserpine, an antipsychotic agent, acts by depleting biogenic amines from vesicular storage sites (Bertler et al. 1956; Shore et al. 1955). The drop in transmitter available for release and postsynaptic action has been correlated with the drug's antipsychotic efficacy and it has been suggested that neuronal overactivity might cause psychosis. In particular, the acute paranoid states of schizophrenia with hallucinations, delirium and delusions have been correlated with a neuronal hyperactivity.

α-Methyl-p-tyrosine, a potent blocker of tyrosine hydroxylase activity, has not been found to be effective in the treatment of schizophrenia (Gershon et al. 1967), although it can potentiate the antipsychotic effects of neuroleptics (Wålinder et al. 1976). Together with the disappearance or reduction of psychosis after discontinuation or reduction of antiparkinsonian drugs, these data implicate a catecholaminergic overactivity in states of paranoid psychosis. However, administration of α-methyldopa, causing lowering of brain dopamine by inhibition of dopa decarboxylase, is ineffective against chronic schizophrenia (Pecknold et al. 1972).

Neuroleptics antagonize the action of dopamine at receptor sites and, consequently, by increasing the firing rate of dopamine neurons accelerate dopamine turnover. As such, neuroleptics are potent antipsychotics but do not influence negative symptoms of schizophrenia. Therefore, a dopamine, or noradrenaline hypothesis seems to be applicable only to acute psychoses and not to the negative symptoms which were particularly important in Kraepelin's dementia praecox. Neuroleptic pharmacology has, however, provided us with a near definite site of action of these antipsychotic drugs. Their clinical efficacy (as shown by mean clinical antipsychotic dose) correlates extremely well with their antagonist action at the dopamine D_2 receptor site in vitro (Seeman 1980).

Human Postmortem Brain Studies

Despite the enormous difficulties in obtaining adquate postmortem human brain samples, with short postmortem time, case histories and data from pathologists and neuropathologists (Riederer 1983), there is convincing evidence that catecholamine systems are involved in certain symptoms of schizophrenia (Table 1). Thus elevation of ^{3}H-butyrophenone binding sites (dopamine D_2 receptors) by 50%–100% has been noted by the groups of Seeman and Crow (Lee et al. 1978; Owen et al. 1978; Lee and Seeman 1980; Crow et al. 1978) and was later confirmed in a larger series of schizophrenic brain tissue samples (Seeman 1980; Crow et al. 1981). Mackay et al. (1978) did not initially observe any change in adenylate cyclase independent binding sites, although in later, larger studies an increase was found (Mackay et al. 1982).

Our own results obtained from chronic, long-term institutionalized schizophrenics seem not to be in line with the above reports, as neuroleptic-free patients can display a significantly lower number of ^{3}H-spiroperidol binding sites (Reynolds et al. 1980; 1981a). However, most of the neuroleptic-free patients belonged to group 295.6 according to the ICD classification (8th edition), indicating "residual" symptomatology with negative symptoms as more apparent. Patients with acute paranoid psychosis (ICD 295.3) have increased receptor density but were generally treated with neuroleptics until shortly before death (Riederer et al. 1984). Certainly, the residual concentration of neuroleptic interferes with ligand binding (Owen et al. 1979; Reynolds et al. 1980). Neuroleptic treatment can certainly elevate D_2 receptor number in animals (Clow et al. 1950), and we believe this medication to be responsible for the elevation in schizophrenic brain tissue (discussed in the chapter by Reynolds in this volume).

Other parameters which may affect or indicate the state of dopaminergic function have also been studied. Dopamine-sensitive adenylate cyclase activity was unchanged (Carenzi et al. 1975), while Cross et al. (1980) have not found alterations in ^{3}H-cis-flupenthixol-labelled D_1 receptors. More recent data, however, do indicate subtle changes in such receptor units (Memo et al. 1983). No changes have been identified in presynaptic receptor sites as defined by ^{3}H-apomorphine binding (Lee et al. 1978; Lee and Seeman 1980), (1)-6,7-dihydroxy-2-aminotetralin (^{3}H-ADTN) (Cross et al. 1979) and ^{3}H-dopamine (Seeman and Lee 1980). Furthermore, no significant alterations have been detected in MAO-A or -B activity in 14 brain areas (Cross et al. 1977) and in dopamine-β-hydroxylase (DBH) and catechol-o-methyltransferase activity of six brain areas (Cross et al. 1978), although a deficit in DBH activity has been reported (Wise and Stein 1973).

Abnormal increases of dopamine and homovanillic acid have been reported in schizophrenic brain tissue, although such results are not always consistently found (Crow et al. 1981). More recently, some new information has indicated a profound, if strictly circumscribed, abnormality of a dopaminergic system in schizophrenia. Psychological and physiological measurements and the association of schizophreniform psychosis with left-sided temporal lobe epilepsy support the view of schizophrenia as a dysfunction of the left temporal lobe (Flor-Henry 1969). One of the regions of the medial temporal lobe, the amygdala, receives a dopaminergic innervation from the ventral tegmental area. In measuring the dopamine concentration of the left and right amygdalae in postmortem tissue of schizophrenic patients, Reynolds (1983) has found an increase only in the amygdala of the left cerebral hemisphere (compared both to the right hemisphere and a control group). No changes or laterality were observed in dopamine levels of another brain area, the caudate nu-

Table 1. Human Postmortem Brain Findings in Schizophrenia

Transmitter/ metabolite/ enzyme/receptor	Neuroleptic-free	Neuroleptic-treated
Dopamine	CAUD ↑ = PUT ↑ = ACC ↑ = LOA = VSEPT ↑ SPER ↑ AMYG ↑ [a]	CAUD ↑ PUT = ACC =
DOPAC	CAUD = PUT = ACC =	CAUD = PUT = ACC =
HVA	CAUD ↓↑ = PUT = ACC = LOA = VSET =	CAUD ↓ PUT = ACC =
Noradrenaline	CAUD = PUT ↑ = ACC ↑ = LOA = HYP ↑ AMYG = VSEPT ↑ NST ↑ PMT = CMAM ↑	
MHPG	ACC ↑ HYP ↑	
TRP	PUT =	
KYN	PUT =	
Serotonin	CAUD = PUT ↑ = ACC ↑ P GALL ↑ LOA = MAO ↑ AHYP = LHYP ↑	
5-HIAA	CAUD = PUT = ACC ↑ GPALL ↑ LOA = MOA = AHYP = LHYP =	
TH	CAUD = PUT = ACC =	CAUD = PUT = ACC =
DDC	CAUD =	CAUD =
DBH	HYP = HIPP = PCORT = OCORT = TCORT =	HYP = HIPP ↓ PCORT = FCORT = CORT = TCORT =
MAO	CAUD = PUT = ACC = SN = AMYG = THAL = HYP = CING = CER = 4CORT =	CAUD = PUT = ACC = SN = AMYG = THAL = HYP = CING = CER = 4CORT =
COMT	CAUD = PUT = ACC = SN = HIPP = HYP = AMYG = CING = THAL = CER = ACC = THAL = 4CORT =	CAUD = PUT = ACC = SN = HIPP = HYP = AMYG = CING = THAL = CER = 4CORT =
GABA	ACC = THAL =	
GAD	CAUD = PUT = ACC = AMYG =	CAUD = PUT = ACC = AMYG =
CAT	CAUD = AMYG ↑ = HIPP ↑ = CAUD/PUT ↑ =	CAUD = AMYG ↑
MET-ENK	CAUD ↑ PUT = GPALL = ACC = HYP = AMYG ↓	
Substance P	CAUD ↑ PUT = GPALL = ACC = HYP = AMYG ↓	
CCK	HIPP ↓ AMYG ↓ TCORT ↓ CORT =	
Neurotensin	AMYG ↓ =	
VIP	AMYG = CORT =	
Neuropeptide Y	AMYG =	
Somatostatin	AMYG = HIPP ↓ FCORT ↓	
TRH	AMYG =	
Receptors (B_{max})		
Adenylate cyclase	CAUD = ↑ ACC ↑ HIPP = CORT =	
[3]H-Flupentixol	CAUD = ↑ PUT ↑ = FCORT =	
[3]H-Spiroperidol	CAUD ↑ = PUT ↓↑ ACC ↑ = FCORT =	CAUD ↑ PUT ↑ ACC ↑
[3]H-Haloperidol	CAUD ↑ PUT ↑ ACC ↑	
[3]H-ADTN	CAUD = PUT =	
[3]H-Apomorphine	CAUD = PUT = ACC =	
[3]H-Dopamine	CAUD = PUT =	

Table 1. (continued)

Transmitter/ metabolite/ enzyme/receptor	Neuroleptic-free	Neuroleptic-treated
^{3}H-WB 4101 (α)	CAUD = PUT = FCORT =	
^{3}H-Dihydro- alprenolol (β)	CAUD =	
^{3}H-β-Propranolol	PUT =	
^{3}H-GABA	CAUD = PUT =	
^{3}H-Diazepam	CAUD = PUT =	
^{3}H-QNB	CAUD = PUT =	
^{3}H-Serotonin	FCORT = CAUD = PUT =	
^{3}H-LSD	FCORT ↓↑ = CAUD = PUT =	
^{3}H-Naloxone	CAUD ↓ PUT = FCORT =	
^{3}H-Ketanserin	CORT =	

DOPAC, 3,4-dihydroxyphenylacetic acid; HVA, homovanillic acid; MHPG, 3-methoxy-4-hydroxyphenylglycol; TRP, tryptophan; KYN, kynurenine; 5-HIAA, 5-hydroxyindoleacetic acid; TH, tyrosine hydroxylase; DDC, dopa decarboxylase; DBH, dopamine-β-hydroxylase; MAO, monoamine oxidase; COMT, catechol-O-methyltransferase; GABA, gamma-aminobutyric acid; GAD, glutamate decarboxylase, CAT, choline acetyltransferase; MET-ENK, met-enkephalin; ADTN, 2-amino-6,7-dihydroxytetraline; LSD, lysergic acid diethylamide; QNB, quinuclidinyl benzilate; CCK, cholecystokinin; VIP, vasoactive intestinal peptide; TRH, thyrotropin-releasing hormone; ACC, nucleus accumbens; AHYP, anterior hypothalamus; AMYG, amygdaloid nucleus; CAUD, caudate nucleus; CER, cerebellum; CING, cingulate gyrus; CMAM, corpus mamillare; CORT, cortex; F, frontal; GPALL, globus pallidus; HIPP, hippocampus; HYP, hypothalamus; LHYP, lateral hypothalamus; LOA, lateral olfactorium; MOA, medial olfactorium; NST, nucleus striae terminalis; O, occipital; P, parietal; PUT, putamen; PMT, paramedial thalamus; SN, substantia nigra; SPER, substantia perforata; T, temporal; THAL, thalamus; VSEPT, ventral septum; 4CORT, four cortex regions (frontal, occipital, parietal, temporal).
↑ increase; ↓ decrease; = no change
[a] Only in left cerebral hemisphere

cleus. Furthermore, no significant differences were notable in the amygdala content of noradrenaline, indicating the neurochemical and regional specificity of these findings. However, the influence of drug treatment cannot be ruled out. Nevertheless, functional laterality reflected by neurochemical asymmetries seems to be an interesting new approach in evaluating possible pathogenic markers in schizophrenia and other psychiatric diseases.

The possible involvement of noradrenaline in the aetiology of schizophrenia has had substantial support (Hornykiewicz 1982), although the evidence has also received some criticism (Iversen et al. 1983). Direct evidence for a noradrenergic overactivity comes from data showing increased noradrenaline concentration in the nucleus accumbens (Farley et al. 1978; Kleinmann et al. 1981) and some other subcortical structures of the schizophrenic brain (Farley et al. 1978). Furthermore, increased concentrations of 3-methoxy-4-hydroxyphenylglycol have been found in nucleus accumbens (Kleinmann et al. 1981). However, the influence of neuroleptic drugs on these postmortem findings has not been excluded.

One CSF study has indicated that increased noradrenaline in schizophrenics is correlated with neuroleptic medication (Gattaz et al. 1983). Ligand studies have not revealed any changes in ^{3}H-propranolol binding (Reynolds et al. 1981 b), ^{3}H-WB 4101 and ^{3}H-dihydroalprenolol binding (Kleinmann et al. 1981).

Despite probable changes in catecholaminergic neurotransmission, other possible pathogenic sites might include serotonergic dysfunction. Studies of metabolites on the kynurenine and serotonin pathways have not shown differences, nor have measurements of ^{3}H-serotonin and ^{3}H-LSD binding in caudate nucleus and putamen (Crow et al. 1981). However, ^{3}H-LSD binding to cortical tissue has been reported to be decreased in a schizophrenic group (Bennett et al. 1979), although attempts to replicate these results with ^{3}H-LSD (Whittaker et al. 1981) or the more specific ^{3}H-ketanserin (Reynolds et al. 1983) were unsuccessful.

Decreased GABAergic activity has been suggested to underline the dopamine hypothesis of schizophrenia. A reduction of inhibitory GABA inputs to dopamine neurons would enhance dopamine neurotransmission. However, glutamate decarboxylase (GAD) activity (GAD is the enzyme synthesizing GABA from glutamic acid), nor ^{3}H-GABA binding (at 10 nM) nor ^{3}H-benzodiazepine binding at 1.6 nM in caudate nucleus and putamen of 20 schizophrenics showed abnormalities (Crow et al. 1981).

Increased activity of choline acetyltransferase (CAT), the enzyme synthesizing acetylcholine from choline and acetyl coenzyme A, has been reported in the medial amygdaloid nucleus, hippocampus and striatum (McGeer and McGeer 1977), although these data have not been confirmed by others (Bird 1980).

The increasing interest in the function of the neuropeptides has, as with all such new neurochemical developments, provided schizophrenia research with a new approach. Opiates have been suggested as a etiological factors in mental disorders because increased opiate peptides have been measured in CSF of schizophrenics and first results with naloxone, an opiate antagonist, have shown beneficial therapeutic effects in schizophrenic patients (Terenius et al. 1976). However, this finding has been disputed, as has the antischizophrenic efficacy of des-tyr-endorphin (Emrich et al. 1980 for review). Intraventricular administration of β-endorphin in animals produces a dose-dependent effect on motor behaviour, with hyperactivity after low dose and inactivity, rigidity and oral dyskinesias following high doses. This behaviour is different from that observed after haloperidol. Opiates can affect the release of dopamine, but it is unlikely that the site of action of des-tyr-endorphin is the dopamine receptor.

Non-opiate peptides have also been under investigation, particularly in the limbic structures of the temporal lobe. Cholecystokinin (CCK), a peptide which may have functional effects on dopaminergic systems, has been found to be reduced in the hippocampus, amygdala and temporal cortex of postmortem schizophrenic brains (Ferrier et al. 1983), albeit only in Crow's type II subgroup (Crow et al. 1981), which probably best corresponds to ICD 295.6 in our studies: the defect state of schizophrenia where negative symptoms predominate. This has been confirmed for the amygdala (Carruthers et al. 1984) in a preliminary study where deficits were also identified in neurotensin (NT), substance P and met-enkephalin, but not in vasoactive intestinal polypeptide (VIP), neuropeptide Y and somatostatin (SRIF). Ferrier et al. (1983) and Nemeroff et al. (1983) were able to confirm this for SRIF, although the former showed a hippocampal deficit of this peptide in type II schizophrenics, while the latter group could only find a loss in frontal cortex. However, Biggins et al. (1983) and Nemeroff et al. (1983) were also unable to find a significant deficit of NT in the schizophrenic amygdala. Ferrier et al. (1983) quote the changes they identify to be independent of previous neuroleptic medication. Moreover, no changes have been found in the amygdala for the thyrotropin-releasing hormone (Biggins et al. 1983). VIP and CCK were not changed in the cerebral cortex (Perry et al. 1981). Furthermore, unchanged glutamate levels in a variety of brain areas do not support the idea of a glutamatergic dysfunction in schizophrenia (Perry 1982).

These (somewhat confusing) findings, taken with the unilateral abnormality in amygdala dopamine, point to a limbic pathology in schizophrenia which, as discussed above, fits in well with the electrophysiological and other clinical evidence for the "pathological site" of psychotic dysfunction. It is also supported by anatomical findings. Increasingly, CAT scanning and even more modern approaches such as positron emission tomography (PET) scanning are pointing towards morphological abnormalities within the schizophrenic brain. Careful assessments of sizes of various nuclei in a large collection of fixed brain tissue (Vogt collection) yielded results indicating atrophy of several limbic and other subcortical areas, including the amygdala, in schizophrenia. Whether the changes in dopamine and in peptide content reflect a heterogeneous atrophy of the different neuronal systems of the limbic lobe is as yet unclear, but is perhaps not, unlikely. With the trend towards a more careful assessment of schizophrenia, exemplified by Crow's distinction of two syndromes, coupled with the substantial technical advances available in the clinic (such as PET and nuclear magnetic resonance) and in neuroscience (such as receptor autoradiography and other microneurochemical techniques), a greater understanding of the disease need not be far away.

References

Bennett JP, Enna SJ, Bylund DB, Gillin JC, Wyatt RJ, Snyder SH (1979) Neurotransmitter receptors in frontal cortex of schizophrenics. Arch Gen Psychiatry 36:927–934

Bertler A, Carlsson A, Rosengren E (1956) Release by reserpine of catecholamines from rabbits heart. Naturwissenschaften 22:521

Biggins JA, Perry EK, McDermott JR, Smith AI, Perry RH, Edwardson JA (1983) Post mortem levels of thyrotropin-releasing hormone and neurotensin in the amygdala in Alzheimer's disease, schizophrenia and depression. J Neurol Sci 58:117–122

Bird ED (1980) A brain tissue resource center to promote research in schizophrenia. In: Baxter C, Melnechuk R (eds) Perspectives in schizophrenia research. Raven, New York

Bleuler E (1923) Lehrbuch der Psychiatrie. Springer, Berlin

Carenzi A, Gillin JG, Guidotti A, Schwartz MA, Trabucchi M, Wyatt RJ (1975) Dopamine-sensitive adenylate cyclase in human caudate nucleus. A study in control subjects and schizophrenic patients. Arch Gen Psychiatry 32:1056–1059

Carlsson Å (1978) Does dopamine have a role in schizophrenia? Biol Psychiatry 13:3

Carruthers B, Dawbarn D, de Quidt M, Emson PC, Hunter J, Reynolds GP (1984) Changes in neuropeptide content of amygdala in schizophrenia. Br J Pharmacol. in press

Clow A, Theodorou A, Jenner P, Marsden CD (1980) Changes in rat striatal dopamine turnover and receptor activity during one year's neuroleptic administration. Eur J Pharmacol 63:135–144

Cross AJ, Crow TJ, Glover V, Lofthouse R, Owen F, Riley GJ (1977) Monoamine oxidase activity in post mortem brains of schizophrenics and controls. Br J Clin Pharmacol 4:719

Cross AJ, Crow TJ, Killpack WS, Longden A, Owen F, Riley GJ (1978) The activities of brain dopamine-β-hydroxylase and catechol-o-methyl transferase in schizophrenics and controls. Pharmacology 59:117–121

Cross AJ, Crow TJ, Owen F (1979) The use of ADTN (2-amino-6,7-dihydroxy-1,2,3,4-tetrahydronaphthalene) as a ligand for brain dopamine receptors. Br J Pharmacol 64:87–88

Cross A, Crow TJ, Owen F (1980) ³H-cis-Flupenthixol (³H-FPT) binding in post-mortem brains of schizophrenics – evidence for a selective increase in dopamine D-2 receptors. Neuropsychopharmacol 4:147

Crow TJ, Johnstone EC, Longden AJ, Owen F (1978) Dopaminergic mechanisms in schizophrenia: the antipsychotic effect and the disease process. Life Sci 23:563–568

Crow TJ, Owen F, Cross AJ, Ferrier N, Johnstone EC, McCreadie RM, Owens DGC, Poulter M (1981) Neurotransmitter enzymes and receptors in post-mortem brain in schizophrenia: evidence that an increase in D₂ dopamine receptors is associated with the type I syndrome. In: Riederer P, Usdin E (eds) Transmitter biochemistry of human brain tissue. Macmillan, London, pp 85–96

Emrich HM, Zaudig M, Kissling W, Dirlich G, Zerssen D, von Herz A (1980) Des-tyrosyl-γ-endorphin in schizophrenia: a double blind trial in 13 patients. Pharmacopsychiatry 13:290

Farley IJ, Price KS, Hornykiewicz O (1978) Monoaminergic systems in the human limbic brain. In: Livingston KE, Hornykiewicz O (eds) Limbic mechanisms. Plenum, New York, pp 333–349

Ferrier IN, Roberts GW, Crow TJ, Johnstone EC, Owens DGC, Lee YC, O'Shaughnessy D, Adrian TE, Polak JM, Bloom SR (1983) Reduced cholecystokinin-like and somatostatin-like immunoreactivity in limbic lobe is associated with negative symptoms in schizophrenia. Life Sci 33:475–482

Flor-Henry P (1969) Psychosis and temporal lobe epilepsy: a controlled investigation. Epilepsia 10:363–395

Friedhoff AM, Alpert M (1973) A dopaminergic-cholinergic mechanism in production of psychotic symptoms. Biol Psychiatry 6:165–169

Gattaz WF, Riederer P, Reynolds GP, Gattaz D, Beckmann H (1983) Dopamine and noradrenaline in the cerebrospinal fluid of schizophrenic patients. Psychiatry Res 8:243–250

Gershon S, Hekiman LJ, Floyd A, Hollister LE (1967) α-Methyl-p-tyrosine (AMT) in schizophrenia. Psychopharmacology (Berlin) 11:189–194

Hornykiewicz O (1982) Brain catecholamines in schizophrenia – a good case for noradrenaline. Nature 299:484–486

Iversen LL, Reynolds GP, Snyder SH (1983) Pathophysiology of schizophrenia – causal role for dopamine or noradrenaline? Nature 305:577

Janowsky DS, El-Jousef MK, Davis JM (1973a) Provocation of schizophrenic symptoms by intravenous administration of methylphenidate. Arch Gen Psychiatry 28:185–191

Janowsky DS, El-Jousef MK, Davis JM, Sekerke HJ (1973b) Antagonistic effects of physostigmine and methylphenidate in man. Am J Psychiatry 130:1370–1376

Kleinmann JE, Karoum F, Rosenblatt J, Christin Gillin J, Hong J, Bridge TP, Zalcman S, Storch F, Delcarmen R, Wyatt RJ (1981) Catecholamines and peptides in post-mortem schizophrenic brains. In: Perris C, Struwe G, Jansson B (eds) Biological Psychiatry 1981. Elsevier, North Holland pp 711–714

Kraepelin E (1919) Dementia Praecox and paraphrenia. Livingstone, Edinburgh

Lee T, Seeman P (1980) Elevation of brain neuroleptic/dopamine receptors in schizophrenia. Am J Psychiatry 137:191–197

Lee T, Seeman P, Tourtellotte WW, Farley IJ, Hornykiewicz O (1978) Binding of ³H-neuroleptics and ³H-apomorphine in schizophrenic brains. Nature 274:897–900

Luchins D (1975) The dopamine hypothesis of schizophrenia: a critical analysis. Neuropsychology 1:365–378

Mackay AVP, Doble A, Bird ED, Spokes EG, Quik M, Iversen LL (1978) ³H-Spiperone binding in normal and schizophrenic post-mortem human brain. Life Sci 23:527–532

Mackay AVP, Iversen LL, Rossor M, Spokes E, Bird E, Arregui A, Creese I, Snyder SH (1982) Increased brain dopamine and dopamine receptors in schizophrenia. Arch Gen Psychiatry 39:991–997

McGeer PL, McGeer EG (1977) Possible changes in striatal and limbic cholinergic system in schizophrenia. Arch Gen Psychiatry 34:1319

Memo M, Kleinman JE, Hanbauer I (1983) Coupling of dopamine D_1 recognition sites with adenylate cyclase in nuclei accumbens and caudatus of schizophrenics. Science 221:1304–1307

Nemeroff CB, Youngblood WW, Manberg PJ, Prange AJ, Kizer JS (1983) Regional brain concentrations of neuropeptides in Huntington's chorea and schizophrenia. Science 221:972–975

Owen F, Crow TJ, Poulter M, Cross AJ, Longden A, Riley GJ (1978) Increased dopamine-receptor sensitivity in schizophrenia. Lancet 2:223–226

Owen F, Cross AJ, Poulter M, Waddington JL (1979) Change in the characteristics of ^{3}H-spiperone binding to rat striatal membranes after acute chlorpromazine administration: effects of buffer washing of membranes. Life Sci 25:385–390

Pecknold JC, Ananth JV, Ban TA, Lehmann HE (1972) The use of methyldopa in schizophrenia: a review and comparative study. Am J Psychiatry 128:27–31

Perry RH, Dockray GJ, Dimaline R, Perry EK, Blessed G, Tomlinson BE (1981) Neuropeptides in Alzheimer's disease, depression and schizophrenia. A post mortem analysis of vasoactive intestinal peptide and cholecystokinin in cerebral cortex. J Neurol Sci 51 (3):465–72

Perry TL (1982) Normal cerebrospinal fluid and brain glutamate levels in schizophrenia do not support the hypothesis of glutamatergic neuronal dysfunction. Neurosci Lett 28:81–85

Post RM, Fink E, Carpenter WT, Goodwin FK (1975) Cerebrospinal fluid amine metabolites in acute schizophrenia. Arch Gen Psychiatry 32:1063–1069

Reynolds GP (1983) Increased concentrations and lateral asymmetry of amygdala dopamine in schizophrenia. Nature 305:527–529

Reynolds GP, Reynolds LM, Riederer P, Jellinger K, Gabriel E (1980) Dopamine receptors and schizophrenia: drug effect or illness. Lancet 2:1251

Reynolds GP, Riederer P, Jellinger K, Gabriel E (1981a) Dopamine receptors and schizophrenia: the neuroleptic drug problem. Neuropharmacology 20:1319–1320

Reynolds GP, Riederer P, Gabriel E (1981b) Propranolol binding in human brain. Preliminary studies. In: Riederer P, Usdin E (eds) Transmitter biochemistry of human brain tissue. Macmillan, London, pp 105–112

Reynolds GP, Rossor MN, Iversen LL (1983) Preliminary studies of human cortical 5-HT$_2$ receptors and their involvement in schizophrenia and neuroleptic drug action. J Neural Transm [Suppl] 18:273–277

Riederer P (1983) Documentation of biological and procedural variables in human neuroscience. In: Pope A (ed) Human brain dissection. US Dep Health and Human Services, NIH, USA, pp 241–243

Riederer P, Jellinger K, Gabriel E (1984) ^{3}H-Spiperone binding to post mortem human putamen in paranoid and nonparanoid schizophrenics. Proceedings of the VII world congress of psychiatry. Plenum in press

Seeman P (1980) Brain dopamine receptors. Pharmacol Rev 32:229–313

Seeman P, Lee T (1980) Brain dopamine receptors (D_2 and D_3 sites) in Parkinson's disease and schizophrenia. Prog Neuropsychopharmacol 4:609

Shore P, Silver SL, Brodie BB (1955) Interactions of reserpine, serotonin and lysergic acid diethylamine in brain. Science 122:284

Snyder SH (1972) Catecholamines in the brain as mediators of amphetamine psychosis. Arch Gen Psychiatry 27:169–178

Terenius L, Wahlström A, Lindström L, Widerlöv E (1976) Increased CSF levels of endorphines in chronic psychosis. Neurosci Lett 3:157

Tsuang MT (1976) Genetic factors in schizophrenia. In: Grenell RG, Gabay S (eds) Biological foundations of psychiatry. Raven, New York, pp 633–644

Wålinder J, Skott A, Carlsson A, Ross BE (1976) Potentation by methyltyrosine of thioridazine effects in chronic schizophrenics. Arch Gen Psychiatry 33:501

Whitaker PM, Crow TJ, Ferrier IN (1981) Tritiated LSD binding in frontal cortex in schizophrenia. Arch Gen Psychiatry 38:278–280

Wise CD, Stein L (1973) Dopamine-β-hydroxylase deficits in the brain of schizophrenic patients. Science 181:344–347

Biochemical Disturbances in Psychiatric Disorder: Are They Specific to Diagnosis or to Certain Symptoms?

C. M. BANKI, M. ARATÓ, and Z. PAPP

Introduction

Several biochemical markers have been claimed to be helpful in identifying central metabolic alterations that would correlate with some psychiatric disorders. The acid metabolites of monoamine transmitters [5-hydroxyindoleacetic acid (5-HIAA) for serotonin, homovanillic acid (HVA) for dopamine, and 3-methoxy-4-hydroxyphenylglycol (MHPG) for noradrenaline] have frequently been measured in the lumbar cerebrospinal fluid (CSF) of patients as indirect reflections of the turnover of the respective neurotransmitters. There were a number of earlier studies reporting lowered concentrations of one or more of these compounds in the CSF of (mainly depressed) patients, but several others contradicted these; there are many excellent reviews on this topic (Murphy et al. 1978; Van Praag 1977; Annitto and Shopsin 1979; Åsberg and Bertilsson 1979). In fact, recent works have hardly been able to confirm the possibility that any of the usual psychiatric categories have a corresponding specific alteration in the CSF biochemical findings (Vestergaard et al. 1979; Berger et al. 1980; Banki et al. 1981a) or still present conflicting data (Van Praag and De Haan 1980; Gattaz et al. 1982). On the other hand, some physical (Åsberg et al. 1981), psychological (Rydin et al. 1982), electrophysiological (Benson et al. 1983), and specific behavioral, as for example suicidal (Oreland et al. 1981; Banki and Arató 1983), features proved to be correlated with CSF biochemistry. In addition, some "elementary" symptoms, common to more than one clinical syndrome, were also found to be related to certain CSF variables: general psychomotor activity to HVA (Van Praag et al. 1975), anxiety and insomnia to 5-HIAA (Banki 1977), aggression again to 5-HIAA (Brown et al. 1979), and so on. Previously, we found that several separately rated symptoms showed some isolated correlations with one of the two amine metabolites 5-HIAA and HVA, while there was no correlation with the global clinical severity (Banki et al. 1981a). On the basis of similar observations it has been repeatedly proposed that isolated symptoms and behavioral features should be studied for possible biochemical correlations instead of complex nosological entities (Zarifian and Loo 1979; Ågren 1981; Banki 1983). However, most of these studies involved only a very few disorders, and it remained unanswered whether the correlations are present in other conditions.

The serotonin precursor tryptophan (TRY) was sometimes also measured in CSF as an index of brain TRY supply (Young et al. 1976), which in turn is one factor influencing brain TRY synthesis; there was an increase in CSF tryptophan in patients with liver disease (Young et al. 1975) and a correlation, as expected, between

CSF TRY and 5-HIAA, but there was no difference in the mean CSF TRY level between depressed and nondepressed patients (Curzon et al. 1980; Banki et al. 1981 a).

Cortisol in the CSF has become a focus of attention since the demonstration of a hypothalamic dysregulation of cortisol secretion in depressed patients (Doerr 1980; Carroll 1982). Although the dexamethasone suppression test, widely used to detect this neuroendocrine anomaly, did not prove to be as specific for depression as was first suggested (Dewan et al. 1982; Raskind et al. 1982), there were strong indications that it really reflects some type of psychic symptomatology, even if not strictly correspondent to the diagnostic classification. By direct measurement, CSF cortisol was found to be somewhat elevated in depression by Träskman et al. (1980), while neither Coppen et al. (1971) nor Jimerson et al. (1980) could detect any significant difference.

The principal aim of the present work was to study these four CSF biochemical variables in a population of meaningful size consisting of patients with four different clinical disorders, and find correlations between the clinical symptomatology and the CSF data. The main question to be answered was whether there were any possible interdiagnostic differences in these correlations.

Patients and Methods

Cerebrospinal fluid data from 110 female psychiatric inpatients, collected over about 2 years, were analyzed. The age of the subjects ranged from 22 to 69 years (mean 43 ± 11 years), and they had been hospitalized because of actual psychotic symptoms or severely disturbed behavior. Diagnoses were made using DSM-III (Diagnostic and Statistical Manual of Mental Disorders, third edition) criteria: 29 patients suffered from major depression (unipolar or bipolar), 32 had a schizophrenic (or schizophreniform) disorder, 29 were alcohol dependent (with or without a concomitant alcohol withdrawal syndrome or withdrawal delirium), and 20 patients suffered from an adjustment disorder (in ICD-9 terms "reactive" syndromes). All diagnoses were established by the consensus of two expert psychiatrists who used the same semistructured interview on subsequent days. None of the patients included in this study had received neuroleptics, antidepressants, lithium, or meaningful doses of tranquilizers for at least 2 weeks before admission. Individuals with any significant physical, endocrine, or metabolic illness, neurological disease, fever, etc. were carefully excluded, as were those with substance abuse other than alcohol (in the alcohol-dependent group). A concomitant psychiatric disorder (as, e.g., personality disorder, anxiety disorder, somatoform disorder) did not exclude the patient if there was consensus about the primary diagnosis and if the secondary disorders did not include organic mental disorders (except alcohol withdrawal syndromes in the dependent group).

Written informed consent was obtained from all participants, and they remained on placebo for 2–4 days after admission while undergoing laboratory tests and psychopathological rating. Lumbar punctures (LP) were performed at 8.00–9.00 a.m. in a standard setting, after diet restrictions and controlled bed rest (Banki et al. 1981a). CSF samples were stored for no more than 14 days, in the dark and at $-60\,°C$.

5-HIAA, HVA, and TRY were measured by fluorometric procedures, and cortisol was assayed by competitive protein-binding technique as described earlier (Banki and Arató 1983). The 5-HIAA and HVA data were standardized to the mean age and body height of the population because there was significant correlation with both variables as repeatedly reported previously (Åsberg et al. 1981; Banki and Molnar 1981). Neither TRY nor cortisol showed any similar relationship to these variables or to body weight.

Forty-five individual symptoms were separately scored for each patient during the diagnostic interview; the dimensions and the scoring system were closely similar to those employed in the Comprehensive Psychiatric Rating Scale (CPRS) (Åsberg et al. 1978), i.e., we used three defined levels of severity and odd numbers for interpolation. The intraclass correlation with three raters proved to be at least $r = 0.60$ for all items, but it reached $r = 0.75$ for the symptoms analyzed in this study. The selection criterion for the symptoms included was an at least 20% occurrence of ratings greater than 1 in the whole group, in order to avoid markedly skewed distributions and nearly constant variables. In this way we obtained nine symptom dimensions which occurred with sufficient frequency not only in the total group but also in each of four diagnostic subgroups to be analyzed statistically. In addition, the Clinical Global Impression (CGI) scale was completed for all individuals.

Although symptom ratings do not represent continuous variables, parametric statistics are acceptable (Ågren 1981; Mountjoy and Roth 1982). Therefore, we used t-tests, Pearson's correlation coefficients (and tests for their homogeneity using z-transformation), and multiple regression analysis as statistical instruments. All variables in the total group of the 110 patients were symmetrically distributed, and although some symptom variables were not normally distributed we did not perform data transformations in these calculations.

Results

Cerebrospinal fluid mean metabolite levels in the four diagnostic categories are presented in Fig. 1. The only significant difference was the lower 5-HIAA mean concentration in schizophrenia than in major depression. TRY is not given in the figure, but there were no differences in the mean TRY levels in the four groups (4.8 ± 2.3 µmol/l in depression, 4.4 ± 2.0 in schizophrenia, 5.2 ± 2.9 in alcohol dependence, and 4.5 ± 1.8 in adjustment disorder).

There was no significant correlation between the CGI score and any of the four biochemical variables either in the total group or in the subgroups (Table 1); Fig. 2 shows the original 5-HIAA data separately for the diagnoses in relationship with the CGI rating in order to give a visual impression of the situation. The other three metabolites would give similar pictures.

The most important step of the data analysis was to calculate the correlation coefficients for each possible symptom-metabolite pair, both for the total group and separately for the four diagnostic subgroups. This computation yielded 144 coefficients; however, for the sake of clarity we give only the range of the four subgroup correlations and the total r in Table 2. There were five significant coefficients with

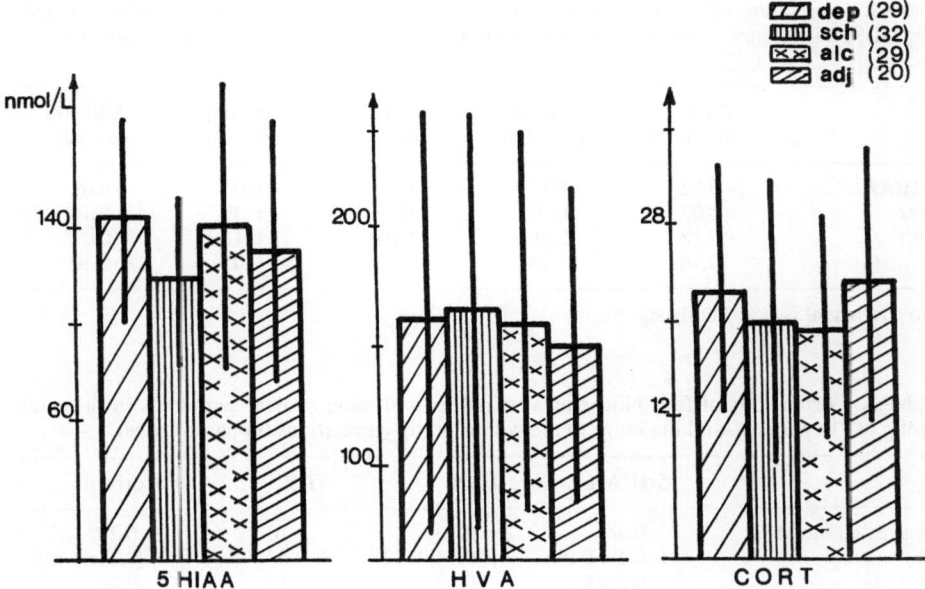

Fig. 1. Means and standard deviations of CSF metabolites in depression (*dep*) schizophrenia, (*Sch*), alcohol dependence (*alc*), and adjustment disorder (*adj*)

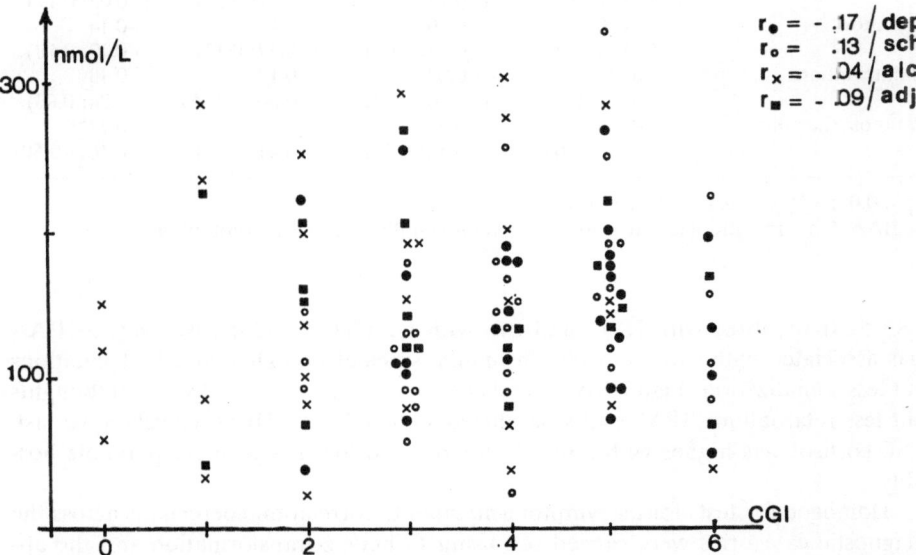

Fig. 2. CSF 5-HIAA and Clinical Global Impression (*CGI*) score in four diagnostic groups: depression (*dep*), schizophrenia (*sch*), alcohol dependence (*alc*), and adjustment disorder (*adj*)

Table 1. Correlation of cerebrospinal fluid 5-HIAA, HVA, TRY, and cortisol with Clinical Global Severity score in depression, schizophrenia, alcohol dependence, and adjustment disorders

	Total ($n=110$)	Depress. ($n=29$)	Schizophr. ($n=32$)	Alc. dep. ($n=29$)	Adjustm. ($n=20$)
5-HIAA	−0.02	−0.17	0.13	−0.04	−0.09
HVA	−0.07	0.03	0.12	−0.09	−0.14
TRY	−0.12	−0.20	−0.06	−0.11	−0.07
Cortisol	0.14	0.28	−0.23	−0.01	0.10

No coefficient is statistically significant

Table 2. Correlations of four biochemical variables with nine clinical symptoms in the total group of 110 patients, and the range of the within-group correlations (in parentheses)

	5-HIAA	HVA	TRY	Cortisol
Depressed mood	0.04 (−0.12; 0.08)	0.10 (−0.14; 0.17)	0.10 (−0.08; 0.10)	0.20* (0.18; 0.40)
Anxiety	−0.27** (−0.23; −0.44)	0.09 (−0.07; 0.32)	0.12 (−0.13; 0.18)	0.06 (−0.14; 0.16)
Somatization	0.20* (0.06; 0.34)	−0.01 (−0.07; 0.13)	0.23* (−0.08; 0.28)	−0.11 (−0.29; 0.05)
Suicidal thoughts	−0.19* (−0.15; −0.43)	0.25** (−0.08; 0.33)	0.08 (−0.18; 0.13)	0.03 (−9.03; 0.32)
Insomnia	−0.37*** (−0.33; 0.54)	0.01 (−0.07; 0.45)	0.22 (−0.18; −0.33)	0.16 (−0.01; 0.42)
Retardation	−0.02 (−0.02; 0.28)	−0.20* (−0.17; −0.52)	−0.10 (−0.11; 0.09)	0.12 (−0.02; 0.17)
Agitation	0.16 (−0.31; 0.29)	0.50*** (0.34; 0.67)	0.14 (−0.17; 0.12)	−0.14 (−0.17; 0.18)
Hallucinations	−0.35*** (−0.41; 0.03)	−0.02 (−0.26; 0.22)	−0.17 (−0.06; −0.35)	−0.13 (−0.24; 0.03)
Paranoia/hostility	−0.18 (−0.33; 0.09)	0.22 (0.18; 0.54)	0.15 (−0.41; 0.20)	−0.27** (−0.16; −0.50)

* $p<0.05$; ** $p<0.01$; *** $p<0.001$
5-HIAA, 5-hydroxyindoleacetic acid; HVA, homovanillic acid; TRY, tryptophan

CSF 5-HIAA, three with HVA, and two with both TRY and cortisol: low 5-HIAA was associated with more anxiety, insomnia, suicidal thoughts, and hallucinations but less somatization; high HVA was related to more agitation and suicidal thoughts but less retardation; TRY results paralleled some of the 5-HIAA correlations; last, CSF cortisol was higher with more dysthymia and lower with more paranoia-hostility.

Homogeneity tests for all symptom-metabolite correlation coefficients across the diagnostic categories were carried out using Fisher's z-transformation and the appropriate χ^2-test. Among the 36 pairs only two proved to be significantly non-homogeneous: the coefficient between 5-HIAA and agitation was −0.31 in depres-

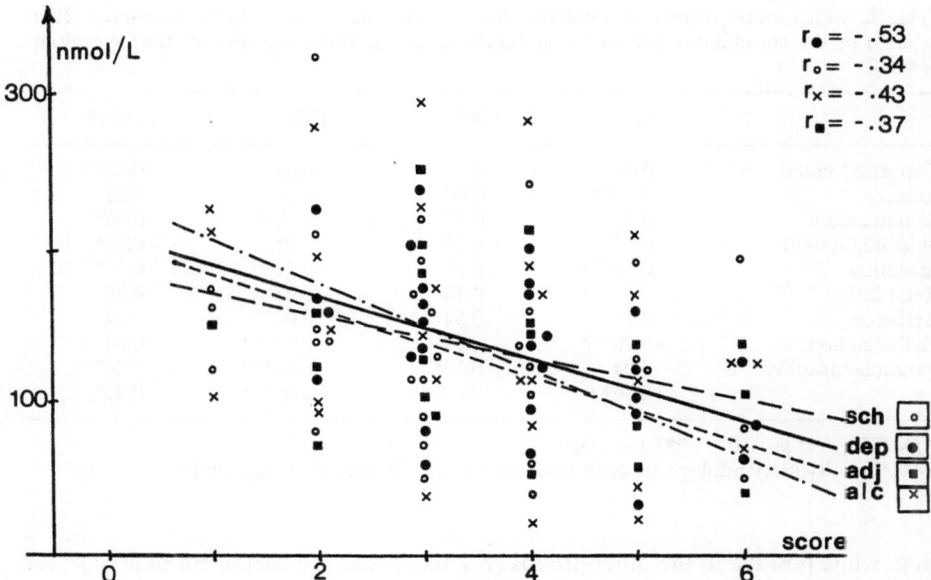

Fig. 3. Relationship between CSF 5-HIAA concentration and the insomnia score in four diagnostic groups: depression (*dep*), schizophrenia (*sch*), alcohol dependence (*alc*), and adjustment disorder (*adj*)

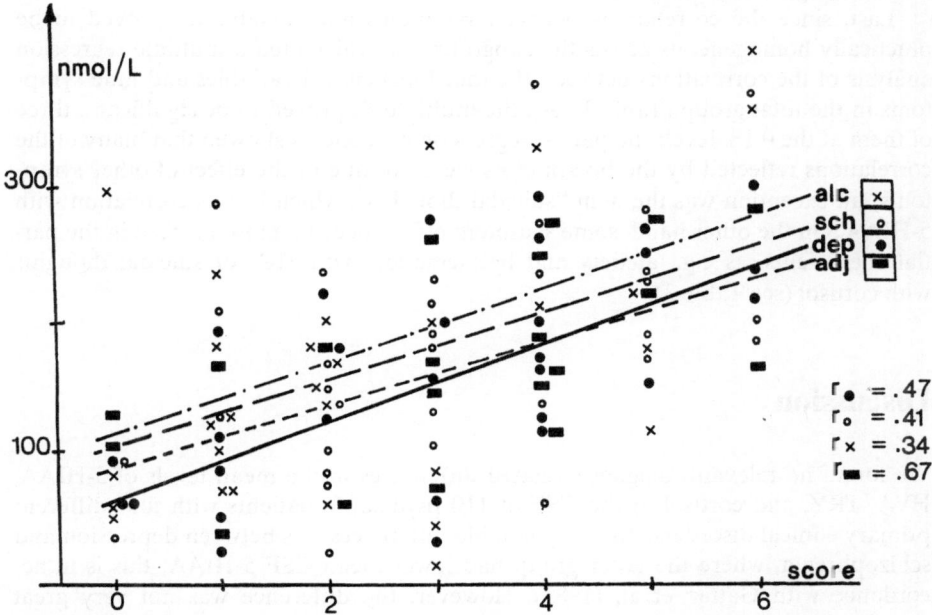

Fig. 4. Relationship between CSF HVA concentration and the agitation score in four diagnostic groups: depression (*dep*), schizophrenia (*sch*), alcohol dependence (*alc*), and adjustment disorder (*adj*)

Table 3. Multivariate regression analysis (beta coefficients and multiple regression coefficients) of the correlations between four biochemical variables and nine clinical symptoms ($n = 110$)

	5-HIAA	HVA	TRY	Cortisol
Depressed mood	0.00	0.07	0.02	0.20**
Anxiety	−0.20**	0.00	0.16*	0.02
Somatization	0.13*	0.00	0.18**	−0.16*
Suicidal thoughts	−0.06	0.29***	0.05	−0.16*
Insomnia	−0.25***	0.00	−0.23**	0.18*
Retardation	0.04	0.10*	−0.05	0.00
Agitation	0.17*	0.54***	0.00	0.04
Hallucinations	−0.26***	−0.09	−0.33***	0.04
Paranoia/hostility	−0.04	0.15*	0.41***	−0.26***
$r =$	0.56***	0.60***	0.49***	0.40*

$* p < 0.05; \quad ** p < 0.01; \quad *** p < 0.001$
5-HIAA, 5-hydroxyindoleacetic acid; HVA, homovanillic acid; TRY, tryptophan

sion, while positive in the other groups ($p < 0.07$), and the coefficient between TRY and paranoia was also negative ($r = -0.42$) in depression while positive in the other groups ($p < 0.06$). Apart from these two values, all the other 34 coefficients were homogeneous in the four clinical groups, and some of them had virtually identical values. Figures 3 and 4 demonstrate two of the strongest correlations and illustrate the interdiagnostic homogeneity.

Last, since the correlations between symptoms and metabolites proved to be practically homogeneous across the subgroups, we calculated a multiple regression analysis of the correlations between the four biochemical variables and nine symptoms in the total group (Table 3). All the multiple r's proved to be significant, three of them at the 0.1% level; the partial regression coefficients showed that many of the correlations reflected by the bivariate r's were not due to the effect of other symptoms; an exception was the item "suicidal thoughts", which lost its correlation with 5-HIAA. On the other hand, some symptom effects became more marked in the partial coefficients, as e.g., anxiety and hallucination with TRY or suicidal thoughts with cortisol (see Table 3).

Discussion

We found no relevant, diagnosis-related differences in the mean levels of 5-HIAA, HVA, TRY, and cortisol in the CSF of 110 psychiatric patients with four different primary clinical disorders. The only notable differences was between depression and schizophrenia, where the latter group had lower mean CSF 5-HIAA; this is in accordance with Gattaz et al. (1982). However, the difference was not very great (118 ± 36 nmol/l in schizophrenia vs 144 ± 44 nmol/l in depression) and there was considerable overlap between the ranges (46–212 nmol/l in schizophrenia and 55–256 nmol/l in depression). This does not seem to justify the use of CSF 5-HIAA as a diagnostic aid, all the less so since the other two diagnostic groups (alcohol dependence and adjustment disorder) had intermediate, strongly overlapping CSF 5-HIAA values.

There were, on the other hand, a number of significant correlations between certain symptoms and CSF metabolite levels. The correlations found by us in a previous group of exclusively depressed female patients (Banki et al. 1981a) were again demonstrated here, in a diagnostically heterogeneous population, and in addition, the group correlations proved to be quite similar across the diagnoses. This seems to allow us to suggest that the concentration of 5-HIAA, HVA, TRY, and cortisol in the CSF may be in fact not so much syndrome- as symptom-related, as first observed by Van Praag et al. (1975) for HVA and hypokinesia.

Age, height, and body weight did not influence the results, since they had either been eliminated by data standardization or were found not to correlate with the biochemical results (Banki and Molnar 1981; Träskman et al. 1980). We searched for some other correlations of the symptoms with several background variables, e.g., duration of illness, number of hospitalizations, episode length before admission, general impairment of adaptation, but could not identify significant associations. The diagnostic groups themselves were also comparable in terms of such variables.

Many of the actual correlations correspond to those found earlier (Ågren 1981; Banki et al. 1981a; Van Praag et al. 1975) or more recently (Gattaz et al. 1982; Oreland et al. 1981); it would be hazardous to try to interpret all significant coefficients. A multiple regression analysis showed that the variance of these CSF biochemical measures can be in a significant degree accounted for by certain symptoms, while they remained largely independent of the global severity, i.e., of the degree of the patients' nonspecific behavioral alteration and/or functional disability. Certainly, CGI is a completely nonspecific measure, but previously we observed a similar lack of relationship between CSF metabolites and total Hamilton score, a measure of depression (Banki et al. 1981b).

The results may, of course, be dependent on the definition of the symptom variables, and this remains necessarily more or less arbitrary since we have no external criteria for deciding which behavioral manifestation constitutes a separate, "elementary" symptom. Our system was a somewhat shortened and modified version of the CPRS (Åsberg et al. 1978), and for the present study some further item concentrations were made to reduce the number of symptom variables. For example, the item "hallucination" comprised commenting voices, other auditory and visual and other hallucinations, and hallucinatory behavior; it may well become clear from a more detailed analysis that these hallucination types would correlate differentially with some CSF biochemical indices.

The data presented in this work could be regarded as an extension and continuation of previous studies from this and other laboratories. Several lines of research seem to lead to the conclusion that central amine metabolic alterations (and their indirect signs, accessible in human studies) are related to certain specific behavioral features, but these do not correspond to diagnostic classification. Not only do suicidal behavior (Oreland et al. 1981), projective test results (Rydin et al. 1982), rapid eye movement sleep parameters (Benson et al. 1983), and other research variables seem to support this statement, but the use of most psychotropic drugs is similarly symptom-oriented and not specifically syndrome-oriented (Goodwin 1977; Beckmann 1981); these drugs are thought to act mainly by altering central neurotransmitter function, so such a symptom-oriented line of CNS biochemical research would be in accord with the everyday experience of psychiatry. Finally, some recent reports indicate that even neuroendocrine parameters such as CSF somatostatin concentration (Rubinow et al. 1983) or thyrotropin-releasing-hormone-induced thyroid-stimulating hormone response (Banki et al. 1983) may be dependent on one or more separate behavioral features, such as sleep or suicide attempts, rather than on the clinical diagnosis. Future work is needed to elucidate the relationship between symptoms and diagnoses as the possible dependent variables of the clinically available biochemical markers of CNS dysfunction.

References

Ågren H (1981) Biological markers in major depressive disorders. Thesis, Uppsala University, Uppsala

Annitto W, Shopsin B (1979) Neuropharmacology of mania. In: Shopsin B (ed) Manic illness. Raven, New York, pp 105–162

Åsberg M, Bertilsson L (1979) Serotonin in depressive illness, studies of CSF 5-HIAA. In: Saletu B, Berner P, Hollister LE (eds) Neuro-psychopharmacology. Pergamon, Oxford, pp 105–115

Åsberg M, Montgomery SA, Perris C, Schalling D, Sedvall G (1978) A comprehensive psychopathological rating scale. Acta Psychiatr Scand [Suppl] 271:5–28

Åsberg M, Bertilsson L, Rydin E, Shalling D, Thorén P, Träskman L (1981) Monoamine metabolites in CSF in relation to depressive illness, suicidal behaviour and personality. In: Angrist B, Burrows GD, Lader M, Lingjoerde O, Sedrau G, Wheatley D (eds) Recent advances in neuro-psychopharmacology. Pergamon, Oxford, pp 257–278

Banki CM (1977) Correlation of anxiety and related symptoms with cerebrospinal fluid 5-HIAA in depressed women. J Neural Transm 41:135–143

Banki CM (1983) Evidence of disturbance of monoamines in depression. In: Van Praag HM, Mendlewicz J (eds) Management of depression with monoamine precursors. Karger, Basel, pp 176–199

Banki CM, Arató M (1983) Amine metabolites and neuroendocrine responses related to depression and suicide. J Affective Disord 5:223–232

Banki CM, Molnar G (1981) The influence of age, height and body weight on cerebrospinal fluid amine metabolites and tryptophan in women. Biol Psychiatry 16:753–762

Banki CM, Molnar G, Vojnik M (1981a) Cerebrospinal fluid amine metabolites, tryptophan and clinical parameters in depression. J Affective Disord 3:91–99

Banki CM, Molnar G, Fekete I (1981b) Correlation of individual symptoms and other clinical variables with CSF amine metabolites and tryptophan in depression. Arch Psychiatr Nervenkr 229:345–353

Banki CM, Arató M, Papp Z, Kurcz M (1983) Cerebrospinal fluid amine metabolites and neuroendocrine findings, biochemical markers in suicidal patients? VII World congress of psychiatry, 11–16 July, Vienna

Beckmann H (1981) Die medikamentöse Therapie der Depression. Nervenarzt 52:135–146

Benson KL, Zarcone VP, Faull KF, Barchas JD, Berger PA (1983) REM sleep eye movement activity and CSF concentration of 5-HIAA in psychiatric patients. Psychiatry Res 8:73–78

Berger PA, Faull KF, Kilkowski J, Anderson PJ, Kraemer H, Davis KL, Barchas JD (1980) Cerebrospinal fluid monoamine metabolites in depression and schizophrenia. Am J Psychiatry 137:174–180

Brown GL, Ballenger JC, Minichiello MD, Goodwin FK (1979) Human aggression and its relationship to cerebrospinal fluid 5-HIAA, MHPG and HVA. In: Sandler M (ed) Psychopharmacology of aggression. Raven, New York, pp 131–148

Carroll BJ (1982) The dexamethasone suppression test for melancholia. Br J Psychiatry 140:292–304

Coppen A, Brooksbank BWL, Noguera R (1971) Cortisol in the cerebrospinal fluid of patients suffering from affective disorders. J Neurol Neurosurg Psychiatr 34:432–435

Curzon G, Kantamaneni BD, Van Boxel P, Gillman PK (1980) Substances related to serotonin in plasma and in lumbar and in ventricular fluids of psychiatric patients. Acta Psychiatr Scand [Suppl] 280:3–20

Dewan MJ, Pandurangi AK, Boucher ML, Levy BF, Major LF (1982) Abnormal DST results in chronic schizophrenic patients. Am J Psychiatry 139:1501–1503

Doerr P (1980) Über die Beziehung zwischen Stimmungsänderung und Aktivität der Hypothalamus-Hypophysen-Nebennierenrinden-Achse bei einem Patienten mit einer in 48stündigen Zyklen verlaufenden Depression. Nervenarzt 51:668–671

Gattaz WF, Waldmeier P, Beckmann H (1982) Cerebrospinal fluid monoamine metabolites in schizophrenic patients. Acta Psychiatr Scand 66:350–360

Goodwin FK (1977) Drug treatment of affective disorders, general principles. In: Jarvik ME (ed) Psychopharmacology in the practice of medicine. Appleton Century-Crofts, New York, pp 241–256

Jimerson DC, Post RM, Van Kammen DP, Skyler JS, Brown GL, Bunney WE (1980) Cerebrospinal fluid cortisol levels in depression and schizophrenia. Am J Psychiatry 137:979–980

Mountjoy CQ, Roth M (1982) Studies in the relationship between depressive disorder and anxiety states. J Affective Disord 4:127–147

Murphy DL, Campbell JL, Costa E (1978) Current status of the indoleamine hypothesis of affective disorders. In: Lipton MA, DiMascio A, Killam KF (eds) Psychopharmacology, a generation of progress. Raven, Oxford, pp 1235–1247

Oreland L, Wiberg Å, Åsberg M, Träskman L, Sjöstrand L, Thorén P, Bertilsson L, Tybring G (1981) Platelet MAO activity and monoamine metabolites in CSF in depressed and suicidal patients and in healthy control. Psychiatry Res 4:21–29

Raskind M, Peskind E, Rivard MF, Veith R, Barnes R (1982) Dexamethasone suppression test and cortisol circadian rhythm in primary degenerative dementia. Am J Psychiatry 139:1468–1471

Rubinow DR, Gold PW, Post RM, Ballenger JD, Cowdry R, Bollinger J, Reichlin S (1983) CSF somatostatin in affective illness. Arch Gen Psychiatry 40:409–412

Rydin E, Schalling D, Åsberg M (1982) Rorschach ratings in depressed and suicidal patients with low levels of 5-hydroxyindoleacetic acid in cerebrospinal fluid. Psychiatry Res 7:229–243

Träskman L, Tybring G, Åsberg M, Bertilsson L, Lantto O, Schalling D (1980) Cortisol in the CSF of depressed and suicidal patients. Arch Gen Psychiatry 37:761–767

Van Praag HM (1977) Depression and schizophrenia, a contribution on their chemical pathologies. Spectrum, New York

Van Praag HM, De Haan S (1980) Central serotonin deficiency – a factor which increases depression vulnerability? Acta Psychiatr Scand [Suppl] 280:89–95

Van Praag HM, Lakke JPW, Schut T (1975) Dopamine metabolism in depression, psychoses, Parkinson disease: the problem of specificity of biological variables in behavioural disorders. Psychol Med 5:138–146

Vestergaard P, Sørensen T, Hoppe E (1979) Biogenic amine metabolites in CSF of patients with affective disorder. In: Saletu B, Berner P, Hollister LE (eds) Neuro-psychopharmacology. Pergamon, Oxford, pp 163–171

Young SN, Lal S, Sourkes S (1975) Relationship between tryptophan in serum and cerebrospinal fluid and 5-HIAA in CSF of man, effect of cirrhosis of liver and probenecid. J Neurol Neurosurg Psychiatry 38:322–331

Young SN, Etienne P, Sourkes TL (1976) The relationship between rat brain and cisternal CSF tryptophan concentration. J Neurol Neurosurg Psychiatry 39:293–303

Zarifian E, Loo H (1979) General survey of the possible correlations between psychiatric diagnoses and biological modifications. In: Saletu B, Berner P, Hollister LE (eds) Neuro-psychopharmacology. Pergamon, Oxford, pp 277–281

Further Studies on the Tyramine Conjugation Deficit in Depressive Illness

M. SANDLER and S. M. BONHAM CARTER

Careful clinical and biochemical scrutiny in recent years has made the conclusion inescapable that depressive illness is a manifestation of somatic disease and probably of more than one. To match the clinical stigmata, biochemical markers are increasingly being uncovered, enabling us to identify and, in some cases, quantify the extent of the biochemical lesion. The validity of some of these markers has yet to be confirmed, but others have now been well authenticated; among their number are decreased platelet 5-hydroxytryptamine uptake (Coppen et al. 1978; Tuomisto et al. 1979), alteration in α_2-adrenoceptor density (Charney et al. 1981), altered ^{3}H-imipramine binding (Langer et al. 1981) and abnormalities in the dexamethasone suppression test (Carroll et al. 1976). And to supplement their number, my colleagues and I have identified a deficit of tyramine conjugation which acts, we believe, as a trait marker in depressive illness.

Our delineation of the tyramine conjugation deficit emerged gradually over a number of years. When we identified a group of depressives who responded specifically to monoamine oxidase (MAO) inhibiting drugs a quarter of a century ago (Pare and Sandler 1959), we speculated that the illness might in some way derive from an overaction of the MAO enzyme system. Much later, during an investigation of patients with dietary migraine, a condition in which Hanington (1976) claimed that tyramine acted as a headache trigger, we noted that many patients belonging to this diagnostic group appeared to have a deficit in the tyramine conjugation pathway (Youdim et al. 1971). Smith et al. (1971) reported similar findings. In the light of our later findings, it is not inconceivable that our (self-selected) patient group had some degree of depressive illness – the term "masked depression" has been applied to a group of patients with headache as a depressive equivalent. Whatever the clinical status of that particular group, the findings led us to further experiments which resulted in our identification of a highly significant decrease in tyramine-conjugating ability in a group of patients with depressive illness (Sandler et al. 1975); this led us naturally back to our old hypothesis of a generalized increase in MAO activity with resulting decrease in transmitter monoamine at the synaptic cleft. But even our subsequent direct demonstration of increased platelet MAO activity in depressive illness (Reveley et al. 1981) does not wholly solve the problem, because the platelet contains solely MAO-B (Donnelly and Murphy 1977). We still have no readily available direct probe of in vivo MAO-A activity.

We followed this first investigation in depressive patients with observations on another more severely afflicted group, this time of medication-resistant depressives (Bonham Carter et al. 1978a), and again noted a highly significant decrease in tyramine conjugating ability. Although there are certain features suggestive of delayed gastrointestinal absorption in depressive illness, indirect experiments (Bonham Carter et al. 1978 b) failed to pinpoint any gut malfunction; nor was there any evidence of sulphate deficiency (Bonham Carter et al. 1980a). What emerged clearly from the investigation of this particular group was that the tyramine conjugation deficit persists even after clinical recovery (Bonham Carter et al. 1978 a). This was the first indication we had that the biochemical lesion was a trait- rather than a state-marker and, as such, a possible predictor of vulnerability to depression. Accordingly, we mounted a prospective study of the phenomenon (Bonham Carter et al. 1980 b).

One manifestation of depressive illness, postpartum depression (as distinct from the "four-day blues"), may well have a disruptive effect on the mother-child relationship, leading to an increase in infant morbidity, if not mortality (H. Caplan, manuscript in preparation). It may thus be extremely important to try to identify prospectively those women who are more likely to develop the illness than others. We therefore screened a group of women attending an antenatal clinic to see whether our oral tyramine challenge test (the administration of 100 mg tyramine orally, followed by the measurement of tyramine sulphate output over a timed period) would identify those more likely to become depressed after childbirth. A lowered urinary output of conjugated tyramine might then point to the need for appropriate social and psychiatric support. This pilot study (Bonham Carter et al. 1980 b) showed clearly that women with low conjugated tyramine output had a significantly higher lifetime incidence of depressive illness compared with those manifesting high output values. None of these patients was depressed at the time of tyramine loading, although two subsequently developed depressive illness in the puerperium of this particular pregnancy.

This was a pilot study and was performed on a series of 70 pregnant women only; a much larger survey of this type urgently needs to be performed. At the present time, the test is tedious to administer both for the clinician and the patient. An obvious modification would be to measure plasma concentration of conjugated tyramine with time, a "tolerance test", and such an approach is now being investigated. Whether the conjugation deficit, whatever its mechanism of production, is specific for tyramine or applies also to other monoamines is still not clear. It has been known for many years that conjugation is an important degradation pathway for orally administered monoamines (Richter 1940; Richter and MacIntosh 1941; Beyer and Shapiro 1945; Häggendal 1963; Resnick 1963; Hengstmann et al. 1975). Thus 10%–15% of orally administered tyramine is so metabolized in normal man (Youdim et al. 1971), whilst other monoamines such as isoprenaline (Morgan et al. 1969) or m-octopamine (Hengstamm et al. 1975) are conjugated to a greater extent. Although our observations are preliminary, our findings (in preparation) would indicate that both p-octopamine and dopamine sulphation are impaired in a manner similar to tyramine conjugation, although the time curve of gastrointestinal absorption of these two amines is somewhat longer than that of tyramine, so that the deficit was almost overlooked. Pin pointing the mechanisms of the tyramine conjugation deficit has also proved to be peculiarly difficult. To test the original hypothesis that an increase in monoamine oxidase activity is involved would require a ready and direct source of MAO-A to supplement that of platelet MAO-B. As mentioned above, such a source has not so far become available. In the near future, we hope to obtain gut biopsy material from depressed patients in remission who are able to give informed consent. The mucosa of the small gut possesses high activity of MAO-A (Squires 1972). However, another theoretical cause of deficient sulphation is obviously some defect of the conjugating enzyme phenolsulphotransferase (PST).

There have recently been a number of important advances in our knowledge of this enzyme (see Sandler and Usdin 1981). Identification of PST activity in the human platelet (Hart et al. 1979; Anderson and Weinshilboum 1980; Rein et al. 1981) has made direct clinical studies feasible at last. In addition, Rein et al. (1982) have been able to show that PST exists in two forms, which they termed M and P: the monoamines and many of their metabolites are specific substrates for the M form, whilst the P form metabolizes a number of exogenous phenols of various kinds (Bonham Carter et al. 1983; Sodha et al. 1983). The two forms of the enzyme differ in tissue distribution, inhibitor specificity and thermostability, as well as substrate preference (Rein et al. 1982). Despite these considerations, careful investigation has revealed no deficit of either PST M or PST P in platelets from depressive patients (Bonham Carter et al. 1981). Final proof, however, must lie in analysis of the gut biopsy material we hope to obtain shortly, for the best evidence we have indicates that sulphate conjugation of an orally administered monoamine predominantly takes place in the small gut mucosa.

Because adenosine triphosphate is essential for the generation of "active sulphate", from which sulphate is transferred via the action of PST to the phenolic recipient, we also measured the concentration of this substance in plasma from depressives and controls. Again, no deficit could be identified (Bonham Carter et al. 1981). The concentration of active sulphate itself needs to be monitored before this line of enquiry is completed. If this, too, proves to be normal, we shall have to fall back on some explanation involving impairment of sulphate uptake.

There are still many unknowns to this equation. We do not yet know whether all clinical subtypes of affective disorder are equally affected or whether the existence of an affective component in other psychiatric illnesses, such as panic/anxiety disorder and alcoholism, will manifest with a greater or lesser degree of the deficit. We do not even known how large a dose of tyramine is necessary to demonstrate the effect. Although we have assumed that our finding points to a trait-dependent deficit, extensive family studies are still required to support this interpretation. However, it is of great interest, in this context, that a tyramine conjugation deficit in depression has very recently been confirmed by Harrison et al. (1982). They showed that after oral tyramine challenge, both unipolar and bipolar endogenous depressives excrete significantly lower amounts of conjugated tyramine than normals, whereas patients with atypical depression are not significantly different in their tyramine excretion pattern from normal controls. Whatever the mechanism of the metabolic defect we have uncovered turns out to be, its empirical use as a trait marker seems likely to provide valuable information in the future.

References

Anderson RJ, Weinshilboum RM (1980) Phenolsulphotransferase in human tissue: radiochemical enzymatic assay and biochemical properties. Clin Chim Acta 103: 79–90

Beyer KH, Shapiro SH (1945) Excretion of conjugated epinephrine related compounds. Am J Physiol 144:321–330

Bonham Carter S, Sandler M, Goodwin BL, Sepping P, Bridges PK (1978a) Decreased urinary output of tyramine and its metabolites in depression. Br J Psychiatry 132:125–132

Bonham Carter S, Sandler M, Sepping P, Bridges PK (1978b) Decreased conjugated tyramine outputs in depression: gastrointestinal factors. Br J Clin Pharmacol 5:269–272

Bonham Carter SM, Goodwin BL, Sandler M, Gillman PK, Bridges PK (1980a) Decreased conjugated tyramine output in depression: the effect of oral L-cysteine. Br J Clin Pharmacol 10:305–308

Bonham Carter SM, Reveley MA, Sandler M, Dewhurst J, Little BC, Hayworth J, Priest RG (1980b) Decreased urinary output of conjugated tyramine is associated with lifetime vulnerability to depressive illness. Psychiatry Res 3:13–31

Bonham Carter SM, Glover V, Sandler M, Gillman PK, Bridges PK (1981) Human platelet phenolsulphotransferase: separate control of the two forms and activity range. Clin Chim Acta 117:333–344

Bonham Carter SM, Rein G, Glover V, Sandler M, Caldwell J (1983) Human platelet phenolsulphotransferase M and P: substrate specificities and correlation with in vivo sulphoconjugation of paracetamol and salicylamide. Br J Clin Pharmacol 15:323–330

Carroll EJ, Curtis GC, Mendels JM (1976) Neuroendocrine regulation in depression. II Discrimination of depressed from nondepressed patients. Arch Gen Psychiatry 33:1051–1058

Charney DS, Heninger GR, Sternberg DE, Redmond DE, Leckman JF, Maas JW, Roth RH (1981) Presynaptic adrenergic receptor sensitivity in depression. Arch Gen Psychiatry 38:1334–1340

Coppen A, Swade C, Wood K (1978) Platelet 5-hydroxytryptamine accumulation in depressive illness. Clin Chim Acta 87:165–168

Donnelly CH, Murphy DL (1977) Substrate- and inhibitor-related characteristics of human platelet monoamine oxidase. Biochem Pharmacol 26:853–858

Haggendal J (1963) The presence of conjugated adrenaline and noradrenaline in human blood plasma. Acta Physiol Scand 59:255–260

Hanington E (1976) Preliminary report on tyramine headache. Br Med J 2:550–551

Harrison W, Cooper T, Quitkin F, Liebowitz M, McGrath P, Stewart J, Klein D (1982) Tyramine excretion test in depressive illness. Abstract of the ACNP Meeting, Puerto Rico, December 1982, p 56

Hart RF, Renskers KJ, Nelson EB, Roth JA (1979) Localization and characterization of phenol sulphotransferase in human placenta. Life Sci 24:125–130

Hengstmann JH, Konen W, Konen C, Eichelbaum M, Dengler HJ (1975) Bioavailability of m-octopamine in man related to its metabolism. Eur J Clin Pharmacol 8:33–39

Langer SZ, Zarifian E, Briley M, Raisman R, Sechter D (1981) High-affinity binding of ³H-imipramine in brain and platelets and its relevance to the biochemistry of affective disorders. Life Sci 29:211–200

Morgan CD, Ruthven CRJ, Sandler M (1969) The quantitative assessment of isoprenaline metabolism in man. Clin Chim Acta 26:381–386

Pare CMB, Sandler M (1959) A clinical and biochemical study of a trial of iproniazid in the treatment of depression. J Neurol Neurosurg Psychiatry 22:247–251

Rein G, Glover V, Sandler M (1981) Sulphate conjugation of biologically active monoamines and their metabolites by human platelet phenolsulphotransferase. Clin Chim Acta 111:247–256

Rein G, Glover V, Sandler M (1982) Multiple forms of phenolsulphotransferase in human tissues: selective inhibition by dichloronitrophenol. Biochem Pharmacol 31:1893–1897

Resnick O (1963) The metabolism of orally ingested epinephrine in man. Life Sci 9:629–636

Reveley MA, Glover V, Sandler M, Coppen A (1981) Increased platelet monoamine oxidase activity in affective disorders. Psychopharmacology (Berlin) 73:257–260

Richter D (1940) The inactivation of adrenaline in vivo in man. J Physiol (Lond) 98:361–371

Richter D, MacIntosh FC (1941) Adrenaline ester. Am J Physiol 135:1–5

Sandler M, Usdin E (1981) Phenolsulphotransferase in mental health research. Macmillan, Basingstoke

Sandler M, Bonham Carter S, Cuthbert MF, Pare CMB (1975) Is there an increase in monoamine-oxidase activity in depressive illness? Lancet I:1045–1049

Smith I, Kellow AH, Mullen PE, Hanington E (1971) Dietary migraine and tyramine metabolism. Nature 230:246–248

Sodha RJ, Glover V, Sandler M (1983) Phenolsulphotransferase in human placenta. Biochem Pharmacol 32:1655–1657

Squires RF (1972) Multiple forms of monoamine oxidase in intact mitochondria as characterized by selective inhibitors and thermal stability: a comparison of eight mammalian species. In: Costa E, Sandler M (eds) Monoamine oxidases – new vistas. Raven, New York, pp 355–370

Tuomisto J, Tukiainen E, Ahlfors UG (1979) Decreased uptake of 5-hydroxytryptamine in blood platelets from patients with endogenous depression. Psychopharmacology (Berlin) 65:141–148

Youdim MBH, Bonham Carter S, Sandler M, Hanington E, Wilkinson M (1971) Conjugation defect in tyramine-sensitive migraine. Nature 230:127–128

Disturbances in Serine-Glycine Metabolism in Relation to Acute Psychoses with Psychedelic Symptoms

J. BRUINVELS and L. PEPPLINKHUIZEN

Introduction

The idea that disease entities might exist in psychiatry can be strongly supported by demonstrating that the disease is linked with a specific pathochemical marker. In general, pathochemical markers comprise the presence or (relative) absence of a biological substance, such as an enzyme, which is only found in latently or manifestly afflicted subjects. With respect to the present findings we wish to extend this narrow definition to include a specific reaction of a patient to an administered substance. Such a procedure can be compared to skin reactions of allergic subjects to hypodermically injected substances, or perhaps more appropriately to the panic reactions of patients with anxiety neurosis to infusions of lactate (Appleby et al. 1981). Demonstrating that the administered substance is pathogenetically related to the diseased state may greatly improve the understanding of the biological basis of the disease, while the specific reaction of the subject to the substance makes a more homogeneous classification of this disorder possible.

Experiences with a number of patients who suffered from acute psychotic episodes characterized by multiple sensory perceptual disturbances – similar to those that occur in LSD- or mescaline-induced states – have opened up new perspectives with regard to the theory that an endogenous synthesis of hallucinogenic substances might be the cause of a psychotic illness. The specific mental reactions evoked in patients after loading experiments with serine or glycine, performed in order to provide evidence for our hypothesis, can also be used as a pathochemical marker for the psychosis studied (Pepplinkhuizen et al. 1980). The case history of the first patient and the ideas that led us to the discovery of this type of pathochemical marking are given in detail below.

Case History

The first patient, 24 years old, unmarried, and female, was admitted to the university hospital after she had attempted suicide by jumping out of a window at the command of her auditory hallucinations. As a result she had broken a leg, which made an operation under general anesthesia necessary. Afterwards she seemed confused: psychiatric examination showed that she was still suffering from auditive (imperative) and also from multicolored visual hallucinations (in the form of "ghosts"). She stated that time was going very fast, was "bombing her," or had come to a complete standstill. Movements were so fast that she was unable to follow them. The white hospital room seemed larger than it was and full of colors, and in combination with the "ghosts" it resembled a temple. Faces of doctors and nurses were impossible to recognize as a result of deformation and coloring.

She was often extremely perplexed and complained about strange bodily feelings and sometimes haptic hallucinations. Anxiety was overwhelming, though she remained passive most of the time. Occasionally, this akinetic state was disrupted by the risk of violent suicide. Voices ordered her to burn or otherwise mutilate herself. Wide pupils, perspiration, tachycardia, etc., even when she was not anxious, indicated a strong orthosympathetic involvement. This state, resembling that described for experimental psychoses induced by hallucinogenic chemicals such as LSD or mescaline, was not affected by neuroleptics.

During the 2 months following admission, neurological symptoms became prominent: there were signs of neuropathy, diplopia, and progressive paresis. Vitamin B_{12} was given in large doses and the neurological symptoms disappeared within a few days. The hallucinations gradually disappeared, but accidentally an analgetic preparation was administered, also containing phenobarbital. Within an hour the hallucinosis fully reappeared. An acute intermittent porphyria (AIP) was then strongly suspected. Cataleptic states lasting several hours and sometimes preceded by abdominal cramps completed the picture. Once an epileptic fit occurred. Fatal outcome was feared when Cheyne-Stokes breathing patterns were observed.

On the assumption that our patient was suffering from AIP, a diet rich in carbohydrates, free from fat, and with minimal protein (especially with a low content of the amino acids methionine, serine, glycine, and tryptophan) was given, when necessary by force. To the surprise of everyone the patient recovered completely within 4 days and no residual symptoms could be demonstrated. Short relapses occurred after the (forbidden) eating of herrings and later of French fried potatoes with mayonnaise, and often in the premenstrual period. Further premenstrual induction of (pre)psychotic symptoms was succesfully prevented by regular medrogestone (Colpro) intake.

Previous Psychotic Episodes

From descriptions of previous clinical admissions the great diversity of psychotic states was striking, not only with regard to the different pictures reported, but also with regard to the length of the psychotic episodes, which varied from days to months, although complete recovery after every episode was evident.

Extreme states of depersonalization, especially of the somatopsychic type, during which the patient said she felt like a robot, as though she was walking on fluff, etc., were reported. These feelings of depersonalization were often followed by auditory hallucinations such as the singing of birds and voices commanding her to commit suicide. Short perceptual distortions of faces and hallucinations of colored stains and circles were also reported. Often her experiences were extremely terrifying – auditory and visual hallucinations of yelping wolves, roaring lions, and deformed human faces and bodies. Hypomanic periods, especially during the evening and at night, and prolonged depressive episodes with distinct feelings of insufficiency and worthlessness were frequently reported, but often these symptoms were dominated by paranoid ideation.

Diagnoses of temporal lobe epilepsy and manic-depressive illness were considered. Electroencephalography did not substantiate the idea of temporal lobe epilepsy. Carbamazepine treatment seemed to worsen the psychosis. Treatment with lithium did not prove to be effective. Neuroleptics had some beneficial effects when sedation occurred. Recovery seemed to occur spontaneously and behavior between psychotic episodes is described as "impressively" normal.

Laboratory Findings

Renal and liver function tests, red and white blood cell count, and hematogram were normal. Vitamin deficiences of folic acid, B_{12} (estimated before the vitamin B_{12} administration), B_1, and B_6 could not be demonstrated. Urinary corticosteroid excretion and thyroid function were also normal. Concentrations of uro- and coproporphyrins in urine and copro- and protoporphyrins in feces were repeatedly normal. After the Dolviran tablet had accidentally been administered, porphobilinogen (PBG) was qualitatively present in urine. Uroporphyrin I-synthetase concentration in red blood cells was also normal. During psychotic states the glycine concentration in urine was increased (Pepplinkhuizen et al. 1980).

Porphyria and Psychoses

The assumption that the underlying basis of the psychoses of our patient was porphyria could not be confirmed biochemically. However, shortly after the successful treatment of the first patient, another girl was admitted to the university hospital who was also suffering from a psychosis characterized by multiple disturbances of sensory perception and catatonic features, acutely evoked after administration of barbiturates and sulfonamide preparations. Abnormal porphyrin excretion was found in urine. Together with the observations mentioned above, this aroused our interest in the biochemical basis of porphyria (Pepplinkhuizen et al. 1980).

Porphyrins are synthesized from succinate and glycine via the enzyme aminolevulinic acid (ALA) synthetase. ALA synthetase is the rate-limiting step in the porphyrin biosynthesis, and this enzyme can be activated by steroids or barbiturates and can be inhibited by glucose and by the end product heme (Meyer and Schmid 1978). A deficient enzyme in the porphyrin pathway (uroporphyrin I-synthetase in the case of acute intermittent porphyria [AIP]) will cause a decreased production of heme. This leads to a decreased feedback repression and activation of ALA synthetase, thus increasing the formation of precursors of the porphyrins (ALA and PBG). During induction of ALA synthetase more succinate and more glycine will be needed. This knowledge does not resolve the problem of how such a psychotic state so similar to one resulting from hallucinogens could be caused by the deranged biochemistry of AIP.

Endogenous Synthesis of Hallucinogenic Agents

Some major theories of endogenous formation of hallucinogenic substances have been reviewed by Bruinvels (1975). From the data presented in the literature it was concluded that psychotomimetic substances could be synthesized endogenously from the normal monoamines under conditions in which methylene (CH_2) and/or methyl (CH_3) tetrahydrofolic acid (THF) concentrations are increased to such an extent that the normal flow of one-carbon units via the vitamin-B_{12}-dependent enzyme homocysteine methyltransferase (yielding methionine) becomes impaired. One-carbon groups are thus trapped. Under such circumstances abnormal methylation of monoamines may become possible.

Later it was shown that it is not methylation but cyclization of monoamines by a nonenzymatic reaction of formaldehyde derived from methylene tetrahydrofolic acid (CH_2 THF) that is probably responsible for the formation of psychotogenic substances (for a review and for the predecessor of this theory, the transmethylation theory of schizophrenia, see Bruinvels et al. 1980; Lewis 1980). On realizing that an excess synthesis of methylated folates could be the key problem, we sought for a biochemical explanation of the case presented.

As pointed out by Bruinvels (1975), the ultimate source of one-carbon neogenesis is the hydroxymethyl group of serine. This amino acid is converted into glycine by the enzyme serine hydroxymethyltransferase (SHMT), in which THF, as a cosubstrate, accepts the one-carbon moiety as methylene (CH_2 THF). The enzyme requires pyridoxal 5'-phosphate as a cofactor (Arnstein and Neuberger 1953; Neuberger 1981). Subsequently, assuming that the patient described was suffering from AIP, we realized that great amounts of glycine had to be formed from serine in order to meet the increased demands for glycine during increased pyrrole synthesis in porphyric attacks. Glycine and succinyl coenzyme A are the precursors in pyrrole synthesis. Thus during excess pyrrole synthesis abundant formation of CH_2 THF will occur, since exogenous sources of glycine are insufficient to meet the daily needs even under normal circumstances (Neuberger 1981).

The first patient was not actually suffering from porphyria, but the hypothesis put forward with respect to porphyria pointed to the fact that any circumstance in which the serine to glycine conversion is increased may be responsible for trapping methylene and methyl tetrahydrofolate. Thus some disturbance in the serine – glycine metabolism was postulated to be the basis of the psychotic episodes.

Among other things, an abnormally increased breakdown of glycine by the cleavage enzyme induces a high turnover, while at the same time the breakdown of glycine again yields a one-carbon group as CH_2THF (Kikuchi 1973). Also, an elevated catabolism of glycine via the succinate-glycine cycle will result in an excess of one-carbon groups (Shemin and Russell 1953). It was argued that in our patient the basic metabolic disturbance is latent during "normal" episodes, but an appropriate stimulus will stimulate serine to glycine conversion to such an extent that methylene trapping will occur.

Therefore, the assumed abnormal catabolism of serine and glycine might be demonstrated even during normal episodes after the patient has been loaded with serine or glycine; plasma serine and glycine concentrations after loading might be different from control values.

Distinct psychopathological changes identical to those previously experienced by the patient during the psychoses emerged after a low oral dose of the amino acids (2 mmol/kg body wt.) but not after glucose, thus suggesting that this was a suitable line for further study. Greatly encouraged by this result we selected more patients for these studies.

Selection

Our first patient had suffered from an acute psychotic disorder characterized by psychedelic symptoms. The loading experiments took place when complete recovery was established. Study of the described psychosis and previous psychotic episodes made clear that even in this patient a diversity of hospital diagnoses had been applied to the psychotic syndromes: oneirophrenia, hallucinosis, hysterical psychosis, schizophreniform psychosis, atypical mania, etc. Nevertheless, in the clinical descriptions of the psychotic episodes sensory perceptual anomalies had been reported, though not emphasized as a special phenomenon, probably due to their very short-lived nature and relative inconspicuousness compared to the often dramatic and florid psychotic content.

Moreover, a retrospective study of 71 case histories of formerly admitted patients who suffered from (episodic) acute psychoses with full recovery and without organic etiology made it clear that of an average of 150 admissions of psychotic patients per year (paranoid, affective, schizophrenic, and unclassifiable psychoses), two or three patients suffered from an acute psychosis with the characteristic anomalies of sensory perception (Pepplinkhuizen et al. 1983).

It also became clear that perceptual anomalies of time sense and a distorted body image were not limited to the psychosis with a disturbed exteroception but were also found in connection with severe depersonalization in other psychoses. In addition to the characteristic perceptual symptoms, affective and so-called schizophrenic symptoms were alternatingly or simultaneously present. Diagnoses such as degeneration, schizophreniform, schizoaffective, and psychogenic psychosis, and atypical mania were made in this group. Other descriptive diagnoses might be: cycloid and atypical psychoses and *bouffée délirante*, diagnoses likely to be dependent on the psychiatrist's country of origin.

According to DSM-III criteria (1980), these psychoses might belong to the "psychotic disorders not elsewhere classified", since the criteria for a schizophrenic, affective, paranoid, or organic disorder were not met. With this knowledge it was possible to select more patients belonging to this unclassifiable group of psychotic disorders with and without the above-mentioned dysperceptions. Schizophrenic patients were also examined, as well as patients suffering from various other conditions (organic and nonorganic), some of them having suffered from sensory perceptual distortions ("dysperceptions" for short) during their psychotic state. Physically and mentally healthy subjects also underwent the same tests.

Procedure

Loading tests were performed on the ward when complete recovery was established (except in the case of the schizophrenic patients). All patients and control subjects were free of psychotropic medication; in three patients lithium therapy was not interrupted, at the request of the patients themselves. Most patients and controls consumed a special diet starting at least 4 days before any loading; the diet was rich in carbohydrate, low in fat, and low in protein, containing a calculated amount of serine (2346 mg) and glycine (1966 mg) (caloric value approx. 2000 kcal or 8400 kJ). Powders containing serine, glycine, and glucose or alanine (the latter two were supposed to be inactive substances) were prepared in a fixed dose of 2 mmol/kg body wt. for each patient. Each powder was administered dissolved in yoghurt half an hour before breakfast and absorption was enhanced by lying the patient in a right-sided position during this half hour. The sequence of administration was unknown to psychiatrist, nursing staff, patients, and control subjects.

The patients were observed by the psychiatrist and nursing staff throughout the rest of the day, with no restrictions on normal daily activities, in order to avoid a laboratory-like experimental setting. In the patients on a carbohydrate-rich diet, hourly blood sampling took place for amino acid determination. Evoked symptoms were assessed by a semistructured interview and completion of the Experiential World Inventory of Osmond and El-Meligi (El Meligi and Osmond 1970).

Clinical Results

In addition to the first patient, 63 other patients and 15 physically and mentally healthy subjects were selected for this study. Thirty patients were expected to react to serine and/or glycine on the strength of the course and clinical picture of their psychotic illness and the reported presence of dysperceptions (the target group). With the other (33) patients such a response was not predicted, even in those (nine) cases with reported dysperceptions. It was deemed or known by the psychiatrists that these nine patients suffered from diseases with another pathogenesis than a disturbed serine-glycine metabolism. Twenty-one patients out of the target group of 30 came up to the expectation that a reaction would occur after intake of serine and/or glycine. One patient also reacted to glucose. Two out of the 33 control patients, who were expected not to react, nevertheless did so, while none of the 15 healthy control subjects reacted. From these numbers it could be concluded that the positive reactions to serine or glycine occurred significantly more in the target group than in the control groups (χ^2 25.0, df 1, $p < 0.0005$, target group versus control patient group). An overview of all the responsive and nonresponsive patients is given below.

Responsive Patients

Altogether 25 patients, 16 females and nine males, showed a psychopathological response starting a few hours after intake of the powder and lasting 2–8 h. Fifteen patients reacted to serine alone, six reacted to glycine, and only two reacted to serine and glycine like the first patient; one showed symptoms after glucose intake. Among these 25 reactive patients were four patients with liver disease, two of them suffering from a porphyric disease. These patients are discussed elsewhere (Pepplinkhuizen et al. 1980, Bruinvels et al. 1980).

The evoked mental reactions differed: no uniform pattern emerged. Most patients reacted with what may be called a more pleasant psychedelic reaction, others with a more unpleasant depersonalization and motor inhibition, and a few with frank psychotic symptoms (see Table 1). One example of a psychedelic reaction (to serine) is given in Appendix A. For all patients and for the observers as well, it was evident that the evoked symptoms were similar to those observed or reported during the "natural" psychosis of the patients.

Some minor signs of apprehension and concomitant vegetative symptoms (tension, sweating, stomach ache, etc.) could be observed or were reported in several patients on loading days. This appeared to be clearly dependent on the inevitable fanciful "information" by other patients and their own exaggerated expectations. Administration of L-alanine in five serine-positive patients and the glucose-positive patient and administration of L-methionine in two serine-positive patients and the glucose-positive patient were without effect, though methionine induced unfavorable strong vegetative symptoms (feeling of sickness, warmth, and perspiration).

Of the 20 patients reactive to serine, glycine, or both who did not suffer from an organic illness, the different hospital diagnoses were: degeneration psychosis, ten; schizoaffective and other psychotic states, three; psychogenic psychosis, five; and schizophreniform psychosis, two. In two out of these 20 patients dysperceptions were so inconspicuous during their natural psychosis that these were not reported, which resulted in inclusion in the group of control patients. During the positive loading test these two patients were overwhelmed by the sudden and severe psychotic outbreak, which was alarming even to observers.

Table 1. Psychopathological reactions after oral loadings

Reaction after loading	Total no. of patients	Dysper- ceptions present	Anom- alous time sense	No. of patients who showed abnormal behavior on			
				Ser- ine	Gly- cine	Both	Pla- cebo [a]
Psychedelic [b]	8	8	5	5	3	0	0
Psychotic [c]	3	3	2	2	0	1	0
Psychedelic and psychotic	7	7	3	4	2	1	0
Extreme depersonalization and motor inhibition	7	4	5	4	1	1	1
Total	25	22	15	15	6	3	1

[a] Glucose or alanine
[b] Dysperceptions, euphoric mood, and ideas of new insight
[c] Frank delusions and/or hallucinations

Nonreactive Patients

From 22 patients out of the total of 39 nonreactive patients such a negative result was to be expected, since no dysperceptions or in a broader sense psychedelic symptoms had been reported. The respective diagnoses and frequencies were: manic-depressive illness, six; unclassifiable psychosis, nine; hysterical psychosis, four; hebephrenia, two; and alcohol hallucinosis, one.

Four patients out of this group of 39 (hysterical psychosis, two; reactive psychosis, one; manic-depressive psychosis, one), however, did complain about oversensitivity to light and noises and/or reported a selective distorted perception of faces, their own or those of others. These patients were erroneously expected to react. It was realized afterwards that these perceptual anomalies were of a very restricted nature, while in the positive patients a more generalized distorted perception was found.

In addition, in five patients with a known organic disorder who had reported dysperceptions, the following diagnoses had been made: variegate porphyria, one; tryptophan malabsorption, one; temporal lobe epilepsy, one; LSD abuse, one; minimal brain damage, one. No response was observed, although the porphyric patient had been predicted to respond.

The remaining eight patients (five schizophrenic patients, two patients suffering from a borderline state, and one patient suffering from a degeneration psychosis) had all reported that dysperceptions were present during their natural psychoses, but were also nonreactive. One schizophrenic patient and one borderline patient had received a false diagnosis before, and hence were wrongly included in the target group. The nonresponsiveness of the patient who suffered from a degeneration psychosis remains unexplained.

In conclusion, it appears that the reactions to serine and glycine are highly specific for the described group of psychoses and support the hypothesis that endogenous synthesis of hallucinogenic substances may be the cause of these psychoses. In some (six out of nine) patients who were wrongly expected to react to serine or glycine, an erroneous diagnosis and/or confusion regarding the character of the dysperceptions appears to be the explanation. In three patients – the glucose-reactive, the porphyric, and the "degenerative" patient – the negative response to serine and glycine remains unexplained.

Differential diagnostic problems likely to be met in the selection of the described psychoses include acute schizophrenic psychoses, temporal lobe epilepsy, and borderline states. With the two patients from the control patient group who unexpectedly reacted strongly to serine and glycine, dysperceptions had been present too briefly and were too unspectacular in relation to the florid, wild psychotic content. Thus any psychosis that meets the criteria for the unclassifiable psychoses according to the DSM-III will have to be scrutinized for the presence of the pathognomonic characteristics, i.e., generalized sensory perceptual distortions of light, colors, taste, shape, distances, etc. (reported to be present at the onset of the psychosis). In addition, the following characteristics, though not pathognomonic, are often found together with the above-mentioned anomalous exteroception: distorted body image, anomalous time sense, and vasovegetative lability.

From the above-mentioned retrospective study and the present results it appears that acute psychoses with dysperceptions and complete recovery have been preferentially diagnosed as "degeneration psychosis." At least in The Netherlands (Hamer 1942; De Leeuw 1946) it is a descriptive diagnosis made when the criteria for schizophrenia and affective, paranoid, and organic disorders are not met and the following characteristics, which do not include dysperceptions, are found:

Acute onset
Complete recovery, often episodic course
Good premorbid functioning
No mental retardation
Clear consciousness, except that perplexity can sometimes be present
Abundant visual hallucinations
Visionary cosmic and ecstatic experiences
Perfunctory delusional ideas (of grandeur)
Motor disturbances
Sudden mood swings, from anxious/depressed to elated/ecstatic
A kaleidoscopic, polymorphous, and alternating clinical picture
(Postpsychotic) depersonalization

It is intriguing to realize that long before the schizoaffective psychosis was discovered (Kasanin 1933), the label of degeneration psychosis was intended to designate exactly all those psychoses bearing resemblance to the functional psychoses but showing their own clinical course. They were also considered to be autochthonous (Kleist 1921; Schröder 1920). This idea, already put forward at the beginning of this century, that the degeneration psychoses form the third group of the so-called functional psychoses, may obtain a renewed and biologically founded basis from improved clinical identification and pathochemical marking.

Changes in Plasma Amino Acids

When the patients had recovered but were still on the carbohydrate-rich diet, venous blood samples were drawn half an hour before breakfast and immediately before the oral intake of serine, glycine, or glucose (2 mmol/kg). Subsequently, blood samples were drawn every full hour after the oral intake of amino acid or glucose uptil the 6th hour. The plasma concentration of amino acids in serine-positive patients was compared to those obtained from a group of healthy controls as well as to plasma concentrations from a small number of glycine-positive and manic-depressed patients. Table 2 shows the plasma concentrations of only those amino acids which were changed in one single diagnostic group of patients. Serine-positive patients showed a lowered plasma level of serine and an increased plasma concentration of taurine, while glycine-positive and manic-depressed patients had a decreased plasma level of alanine and increased plasma level of glycine, respectively. Shea et al. (1981) and Rosenblatt et al. (1982) reported increased glycine concentrations in erythrocytes from manic-depressed patients as a result of lithium treatment. However, none of the three manic-depressed patients in the present study had been treated with lithium.

Table 2. Plasma levels of some amino acids in controls and recovered patients

	$(ser)_{t=0}$	$(gly)_{t=0}$	$(tau)_{t=0}$	$(ala)_{t=0}$
Controls (8)	162 ± 12	236 ± 12	71 ± 5	$318 \pm 23\ \mu mol \cdot l^{-1}$
Serpos (7)	$100 \pm 3*$	217 ± 24	$162 \pm 13*$ (5)	380 ± 73 (5) $\mu mol \cdot l^{-1}$
Glypos (4)	141 ± 14	252 ± 23	75 ± 5 (3)	264 ± 8 (3) $\mu mol \cdot l^{-1}$
M.D. (3)	152 ± 9	$333 \pm 30*$	88 ± 2	$517 \pm 102\ \mu mol \cdot l^{-1}$

Plasma samples were drawn in the morning before oral intake of amino acid and half an hour before breakfast. In brackets the number of patients. Data are expressed as mean $\pm$ SEM.
* $p < 0.05$ vs control group (Student's t-test)
Serpos, patients who showed psychedelic and psychotic symptoms after oral intake of serine; Glypos, patients who reacted after oral intake of glycine; M.D., patients suffering from manic-depressive illness

Table 3. Formation of serine and glycine after oral administration of glycine and serine respectively

Amino acid administered	Controls (8)	Patients	
		Serpos (5)	Glypos (4)
Serine[a]	0.086 ± 0.008 mol gly/mol ser	$0.128 \pm 0.013*$ mol gly/mol ser.	–
Glycine[a]	0.208 ± 0.011 mol ser/mol gly	$0.114 \pm 0.012**$ mol ser/mol gly	0.180 ± 0.013 mol ser/mol gly

Data are expressed as mean $\pm$ SEM
* $p < 0.05$ vs control group (Student's t-test)
** $p < 0.01$ vs control group (Student's t-test)
[a] 2 mmol $\cdot$ kg^{-1} orally
Serpos, patients who showed psychedelic and psychotic symptoms after oral intake of serine; Glypos, patients who reacted after oral intake of glycine

The decreased serine plasma concentration in serine-positive patients suggests that more serine has disappeared from the plasma or that more serine has been metabolized. An impaired synthesis of serine may also be responsible for the decreased plasma level of serine in these patients. The hourly measurements of serine or glycine after loading the patient with the amino acid did not indicate a change in the disappearance rate of serine from plasma in serine-positive patients as compared to controls. However, the disappearance rate of glycine was decreased in serine-positive patients as well as in glycine-positive patients (unpublished results), thus excluding an increased disappearance of serine from the plasma as a cause for the lowered plasma level of serine.

The formation of glycine and serine after the administration of serine and glycine respectively was measured. As shown in Table 3, the formation of glycine from serine was increased by 50% while the reaction from glycine to serine was decreased by 50% in serine-positive patients. The conversion of glycine into serine in glycine-positive patients was, however, not significantly different from that in controls. Since the conversion of serine into glycine and vice versa is an equilibrium reaction, a malfunctioning of the enzyme responsible, SHMT, seems unlikely. In fact the increased formation of glycine from serine may be the result of the impaired conversion of glycine into serine which may be caused by a diminished availability of the cosubstrate N^5,N^{10}-methylenetetrahydrofolic acid (CH_2THF). Nevertheless, the decreased formation of serine from glycine might be responsible, at least in part, for the lowered serine concentration in plasma.

Supporting evidence for an impaired conversion of glycine into serine was also obtained by the failure of glycine to increase alanine formation in serine-positive patients (Fig. 1), in contrast to controls and glycine-positive patients. Administration of serine to serine-positive patients resulted in an increased formation of alanine, which was not different from that obtained in healthy controls.

The increased formation of alanine after serine administration must be ascribed to the conversion of serine into hydroxypyruvate, which occurs simultaneously with the conversion of pyruvate into alanine. Besides a decreased formation of serine from glycine in serine-positive patients, the increased formation of taurine in these patients may also contribute to the lowered plasma level of serine. As shown in Fig. 2, the conversion of serine into glycine occurs simultaneously with the conversion of THF into CH_2THF, which is reduced to N^5-methyl tetrahydrofolic acid (CH_3THF). The methyl group of CH_3THF is transferred to homocysteine, forming methionine. Homocysteine can be formed back from methionine via S-adenosylmethionine (SAM) and S-adenosylhomocysteine (SAH). In addition, methionine

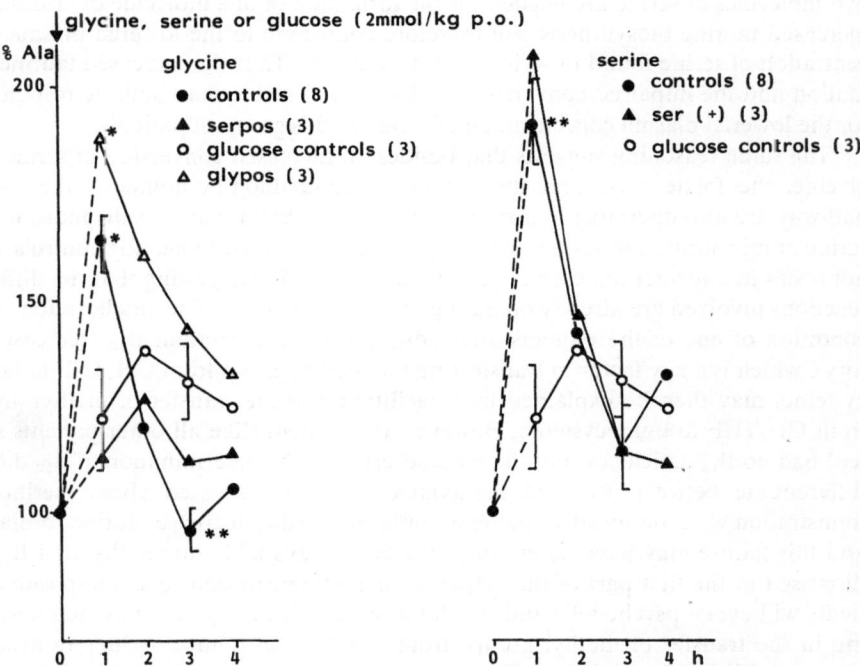

Fig. 1. Conversion of glycine, serine, and glucose into alanine in patients and controls. *serpos,* patients who showed psychedelic or psychotic symptoms after oral intake of serine; *glypos,* patients who reacted after oral intake of glycine. $*P \leq 0.05$, $**P \leq 0.01$ vs t = 0 h

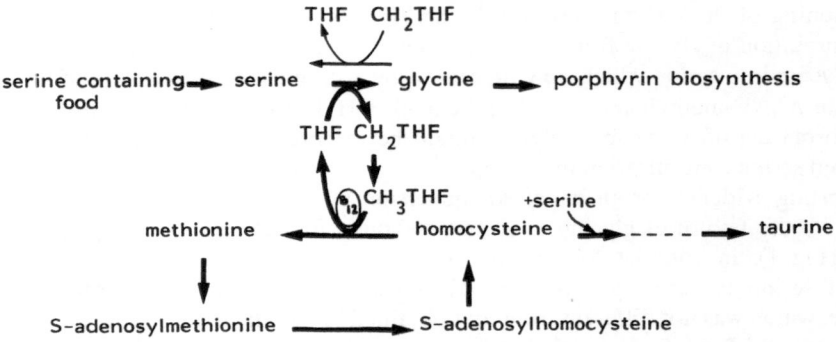

Fig. 2. Pathways involved in the biosynthesis of methyl folates, S-adenosylhomocysteine, and taurine. *THF*, tetrahydrofolic acid; CH_2, methylene; CH_3, methyl

present in the diet (about 1133 mg/day) will also be a source of homocysteine synthesis. Homocysteine will combine with serine forming cystathionine, which will be converted to cysteine, the substrate for the formation of taurine. It is therefore not inconceivable that the increased formation of taurine in serine-positive patients is the result of the shift of the serine-glycine equilibrium to the formation of glycine, with a concomitant increase in CH_2THF as discussed above. In this reaction scheme two molecules of serine are needed for the formation of one molecule of taurine. An increased taurine biosynthesis will therefore contribute to the lowered plasma concentration of serine found in serine-positive patients. Thus the increased taurine formation and the impaired conversion of glycine into serine may both be responsible for the lowered plasma concentration of serine in this group of patients.

The same reasoning suggests that besides an increased conversion of serine into glycine, the folate cycle and the homocysteine-methionine-homocysteine-taurine pathway are also operating at a supranormal level. Preliminary results indicate that serine administration to serine-positive patients, in contrast to healthy controls, does not result in a further increase in plasma taurine levels, suggesting that the different reactions involved are already operating at a maximal level. The amelioration of the condition of one of the patients after administration of vitamin B_{12} (see case history), which is a key factor in transferring the methyl group from CH_3THF to homocysteine, may then be explained by a facilitation of the transfer of methyl groups from CH_3THF to homocysteine. However, this patient (like all other patients studied) had no B_{12} deficiency, but the method used for the determination of B_{12} did not differentiate between B_{12} and methylated B_{12}. As discussed above, serine administration to serine-positive patients could not further increase taurine formation, and this failure may have its origin in the lack of available nonmethylated B_{12}. As discussed in the first part of this paper, serine administration to serine-positive patients will evoke psychedelic and psychotic symptoms. Since B_{12} may be rate-limiting in the transfer of methyl groups from CH_3THF to homocysteine, an accumulation of CH_3THF and CH_2THF may occur as a result of serine administration to these patients increasing the formation of glycine and CH_2THF. As postulated previously (Pepplinkhuizen et al. 1980; Bruinvels et al. 1980), the accumulated CH_2ThF

will decompose into THF and formaldehyde, of which the latter will react with monoamines forming tetrahydroisoquinolines (TIQs) and tetrahydro-β-carbolines (THβCs). Administration of B_{12} in this situation will lessen the accumulated CH_3THF and CH_2THF by binding of the methyl group of CH_3THF.

If glycine instead of serine is administered to serine-positive patients, no mental reaction is evoked and the rise in plasma serine is inhibited by about 50%. It is argued that under these circumstances no sufficient quantity of CH_2THF is available for the conversion of glycine into serine. This explanation fits into the possible mechanism described above, in which most of the CH_3THF, derived from CH_2THF, will be used for the formation of methionine from homocysteine and the final product taurine. Under these circumstances the recovered serine-positive patient will be in a labile situation, since any increase in serine-glycine conversion will result in an extra formation of CH_2THF and CH_3THF, while subsequent transfer of the methyl group to B_{12} will be impossible. Therefore, food containing high concentrations of serine will increase CH_2THF and subsequently increase the formation of TIQs and THβCs. Also a greater demand for glycine, such as an increased porphyrin biosynthesis, will induce an enhancement of the conversion of serine into glycine with the same result.

In conclusion, the described results suggest that in recovered serine-positive patients the decreased plasma level of serine and the increased plasma level of taurine may be used as pathochemical markers. Oral administration of serine to these patients will evoke psychedelic and psychotic symptoms a few hours after administration of the amino acid and may thus be used as a clinical marker for these patients. However, more biochemical experiments are necessary in order to support the mechanism postulated as responsible for the episodic psychedelic and psychotic symptoms in these patients.

Acknowledgments. The authors would like to acknowledge the excellent technical assistance of Y. Dikkerboom and the financial support of Praeventiefonds (project no. 28-753).

Appendix A. Report of Serine-Induced Experiences in a 23-Year-Old Male Patient

At 11 a.m., 2 h after ingestion of the serine powder, the patient spontaneously reported that he was seeing colors more vividly, and that otherwise dull surfaces had become shiny and very beautiful. Colors appeared deeper, more intense. Noises, such as the jingling of keys, were also heard more sharply. A plant appeared as though it were moving: "as if there is streaming in it ... as if I see it moving when it's not." The patient remembered that, during a previous psychotic episode, he thought himself a "reincarnation of Vincent van Gogh," and reported that the present sensations were "just like this experience." He compared the plant to van Gogh's cypresses. This was all very beautiful and meaningful to him. Time seemed to be going faster; he was able to read faster, felt he was understanding more quickly, and was in high spirits.

At 11.30 a.m. he was feeling almighty, tall; he walked through the corridors singing loudly, with a springy, elastic step, looking around with his head held high. His muscles felt more powerful, as though they performed easily. He was making plans for tramping through southern Europe in a black suit to which he had taken a liking for no apparent reason.

Between about 1 p.m. and 2 p.m. the patient became more restless, less cheerful, and more worried. Nevertheless, he felt the urge to draw, and produced 15 psychedelic drawings. He reported that he was now used to the bright colors, but was experiencing new sensations with respect to space: "as if I were in an enormous expanse, as if I had no limits, as if I could go anywhere without any effort," and "it is like when you are at the seaside and look at the horizon, and it is as if the distance between you and the horizon did not exist . . . You are what you see." He also had a strange feeling of extreme close connection, as for example with the observer, and with anything he looked at. Time had now appeared to come to a standstill: it was more like "the clock moving in space." "Twilight would tell me something, but time does not any more." He now felt less of a "gliding" sensation, less that he was "suspended in the cosmos" than he had an hour before.

The effects began to disappear from 2 p.m. onward, and most symptoms had gone by 4 p.m., although a dull feeling (depersonalization) persisted for some hours.

References

Appleby L, Klein DF, Sachar EJ, Levitt M (1981) Biochemical indices of lactate-induced panic: a preliminary report. In: Klein DF, Rabkin J (eds) Anxiety: new research and changing concepts, Raven, New York

Arnstein HRV, Neuberger A (1953) The effect of cobalamin on the quantitative utilization of serine, glycine and formate for the synthesis of choline and methyl groups of methionine. Biochem J 55:259–271

Bruinvels J (1975) Dysmethylation, a possible cause of schizophrenia? In: Van Praag HM (ed) On the origin of schizophrenic psychoses. De Erven Bohn, Amsterdam, pp 30–39

Bruinvels J, Pepplinkhuizen L, Van Tuijl HR, Moleman P, Blom W (1980) Role of serine, glycine, and the tetrahydrofolic acid cycle in schizoaffective psychosis. A hypothesis relating porphyrin biosynthesis and transmethylation. In: Usdin E, Sourkes TL, Youdim MBH (eds) Enzymes and neurotransmitters in mental disease, Wiley, Chichester

De Leeuw CH (1946) Degeneratie-psychoses. In: Horst L van der (ed) Anthropologische Psychiatrie. Van Holkema en Warendorf, Amsterdam, pp 203–264

Diagnostic and statistical manual of mental disorder, DSM III, 3rd edn (1980) Am Psychiatry Assoc, Washington DC

El Meligi AM, Osmond H (1970) Manual for the clinical use of the Experiential World Inventory. Mensana, New York

Hamer BC (1942) Degeneratiepsychose, een katamnestisch onderzoek. Thesis, University of Amsterdam, Nauta, Zutphen

Kasanin J (1933) The acute schizoaffective psychoses. Am J Psychiatry 13:97–126

Kikuchi G (1973) The glycine cleavage system: composition, reaction mechanism and physiological significance. Mol Cell Biochem 1:169–187

Kleist K (1921) Autochtone Degenerationspsychosen. Z Ges Neurol Psychiatr 69:1–11

Lewis ME (1980) Biochemical aspects of schizophrenia. In: Youdim MBH, Lovenberg W, Sharman DF, Lagnado JR (eds) Essays in neurochemistry and neuropharmacology, vol 4. Wiley, Chichester, pp 1–67

Meyer UA, Schmid R (1978) The porphyrias. In: Stanburg JB, Wijngaarden JB, Frederickson DS (eds) The metabolic basis of inherited disease, 4th edn. McGraw-Hill, New York, pp 1166–1220

Neuberger A (1981) The metabolism of glycine and serine. In: Neuberger A, Van Deenen LLM (eds) Comprehensive biochemistry, vol 19 a. Elsevier, Amsterdam, pp 257–303

Pepplinkhuizen L (1983) Disturbances of serine and glycine metabolism as a cause of episodic acute polymorphous psychoses. Thesis, Erasmus University, Rotterdam

Pepplinkhuizen L, Bruinvels J, Blom W, Moleman P (1980) Schizophrenia-like psychosis caused by a metabolic disorder. Lancet I:454–456

Rosenblatt S, Leighton WP, Chanley JD (1982) Elevation of erythrocyte glycine levels during lithium treatment of affective disorders. Psychiatry Res 136:203–214

Schröder K (1920) Degenerativer Irresein und Degenerationspsychosen. Z Ges Neurol Psychiatr 60:119–126

Shea PA, Small JG, Hendire HC (1981) Elevation of choline and glycine in red blood cells of psychiatric patients due to lithium treatment. Biol Psychiatry 16:825–830

Shemin D, Russell CS (1953) Delta-amino levulinic acid, its role in the biosynthesis of porphyrins and purines. J Am Chem Soc 75:4873–4874

Peptides and Amines in Affective Disorders

A. GJERRIS and O. J. RAFAELSEN

Modern psychopharmacology gave rise to the amine hypotheses of affective disorders, and the search began for amine metabolites that could fulfil our hopes of biological markers of depression and mania. It is no disgrace to study what one is able to study, just as one should only move stones that are not too heavy. With respect to the neurotransmitters, this has for quite some years meant that brain studies were concentrated on three amines, noradrenaline, dopamine and serotonin, and their metabolites. In spite of the intensive and extensive research over two decades, it is our seasoned opinion that little of lasting value has come out of this dedicated struggle. Our own studies are characterized by:

1. Analysis of catecholamines proper and not their metabolites
2. Studies of a series of peptides that presumably function as neuromodulators in the CNS – if not as neurotransmitters in their own right.

Adrenaline has in recent years been recognized as a brain neurotransmitter, quantitatively much more sparsely present in the brain and CSF than noradrenaline (Hökfelt et al. 1974). In collaboration with Christensen of Copenhagen, we have performed two studies on CSF adrenaline in depressed patients. The first study, including only endogenously depressed patients, showed a reduction in CSF adrenaline of some 75%, a reduction which normalized after clinical recovery due to electroconvulsive therapy (Christensen et al. 1980).

The second study included both endogenously and non-endogenously depressed patients (Gjerris et al. 1981), and CSF adrenaline was found to be significantly reduced in both groups compared to controls. In both studies CSF noradrenaline showed no differences between depressed and recovered patients and controls (Figs. 1 and 2).

According to these results we found it valuable to investigate the influence of different antidepressant principles on central adrenaline and noradrenaline. In an animal experimental study we treated rats with isocarboxazide and measured the concentrations of CSF adrenaline and noradrenaline. Adrenaline had doubled after only 24 h and continued to increase, reaching some 250% after 6 weeks of continuous treatment. Rat CSF noradrenaline increased more slowly, and the percentage increase was much smaller, reaching only some 40% after 6 weeks (Gjerris et al. 1984) Figs. 3 and 4).

This is in accordance with the studies of Fuller (Fuller and Hemrick-Leucke 1981) and Da Prada (Da Prada et al. 1983), who, using the new reversible monoamine oxidase inhibitors, found that the increase in adrenaline in rat hypothalamus was faster, more pronounced and longer lasting than the increase in noradrenaline.

Furthermore, Roth et al. (1982) reported that adrenaline in rat hypothalamus is particularly responsive to acute and chronic stress.

Among the brain and CSF peptides we have studied cholecystokinin, thyrotropin-releasing hormone, vasopressin, vasoactive intestinal polypeptide (VIP) and a few others. The only positive finding was with VIP, which was measured by a radioimmunoassay by Fahrenkrug of Copenhagen (Fahrenkrug 1979).

The scatter of results in depressive patients was quite large, but when the Newcastle Rating Scales for Depression and the ICD-8 criteria were applied to distinguish between endogenous and non-endogenous depression, it turned out that the patients with non-endogenous but not those with endogenous depressions had a 50% reduction in CSF VIP (Fig. 5). Furthermore, there was no change in these low

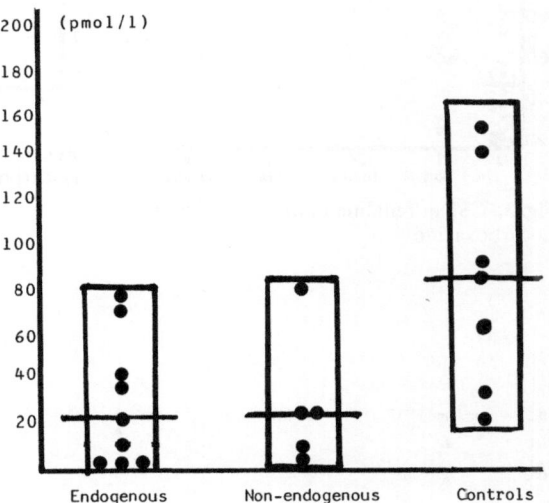

Fig. 1. CSF adrenaline in depression (medians and ranges)

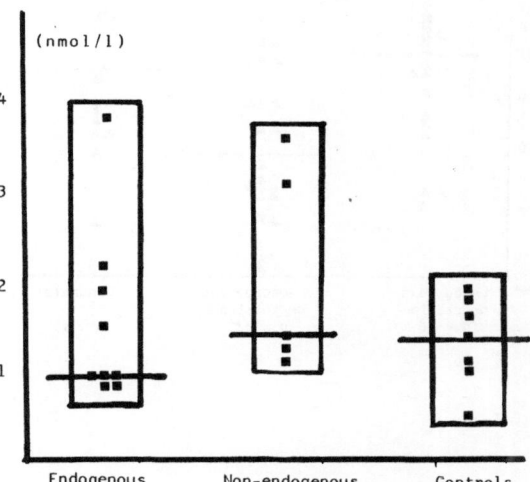

Fig. 2. CSF noradrenaline in depression (medians and ranges)

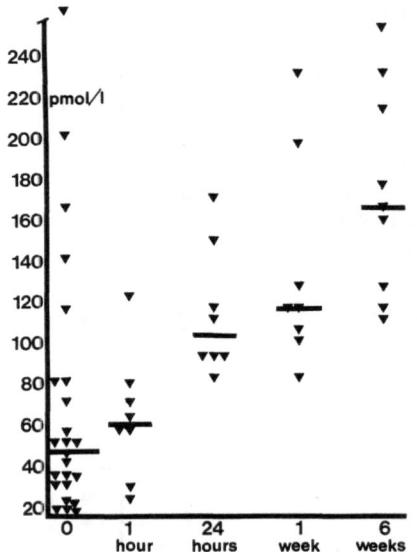

Fig. 3. CSF adrenaline in rats treated with isocarboxazide

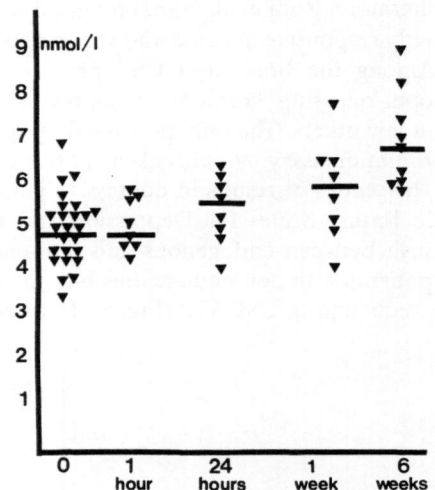

Fig. 4. CSF noradrenaline in rats treated with isocarboxazide

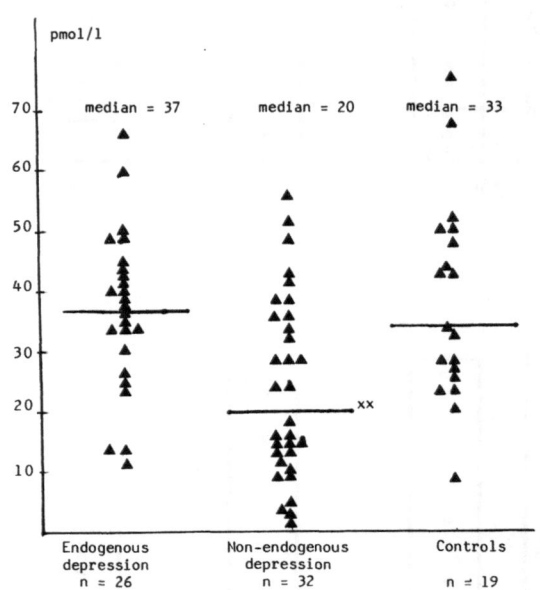

Fig. 5. CSF vasoactive intestinal polypeptide in depression classified according to the Newcastle Rating Scale for Depression 1965. xx, $p \leqq 0.001$

values when the patients had recovered from their depression (Gjerris et al. 1981). We do not know much about the function of VIP in the CNS, but it is supposed to have a vasodilator effect. Moreover, intravenous injection of VIP gives rise to increased serum prolactin levels (Fahrenkrug and Emson 1982). It is noteworthy that VIP is present in CSF in concentrations five to ten times as high as in blood and that it is localized to neurons in relation to the limbic system (Fahrenkrug 1979).

With respect to the coexistence of classical neurotransmitters and peptides in neurons, it has been reported that VIP is found in acetylcholine-containing neurons (Hökfelt et al. 1980). We are now hoping that others will start to replicate our findings, which we at present interpret as two biological markers for depression. One of these (CSF adrenaline) is state dependent (disappearing when the depression disappears and undiscerning between endogenous and non-endogenous depression). The other is state independent and may thus be a trait marker allowing the diagnosis of a reaction potential, a mode for depression, in some individuals; in addition, this marker seems to distinguish between endogenously and non-endogenously depressed patients, and if this is true, it will represent a new biological differentiation that will promote research, diagnosis and treatment in affective disorders (Rafaelsen 1980).

References

Christensen NJ, Vestergaard P, Sørensen T, Rafaelsen OJ (1980) Cerebrospinal fluid adrenaline and noradrenaline in depressed patients. Acta Psychiatr Scand 61:178–182

Da Prada M, Kettler R, Keller HH, Haefely WE (1983) Neurochemical effects in vitro and in vivo of the antidepressant Ro 11-1163, a specific and short-acting MAO-A inhibitor. Mod Probl Pharmacopsychiatry 19:231–245

Fahrenkrug J (1979) Vasoactive intestinal polypeptides. Digestion 19:149

Fahrenkrug J, Emson PC (1982) Vasoactive intestinal polypeptide: functional aspects. Br Med Bull 38:265–270

Fuller RW, Hemrick-Leucke SK (1981) Elevation of epinephrine concentration in rat brain by LY51641, a selective inhibitor of type A monoamine oxidase. Res Commun Chem Pathol Pharmacol 32:207–221

Gjerris A, Jensen E, Christensen NJ, Rafaelsen OJ (1981) Adrenaline and noradrenaline in psychiatric disorders. In: Perris C, Struwe G, Jansson B (eds) Biological psychiatry. Elsevier/North-Holland, pp 565–568

Gjerris A, Barry DI, Christensen NJ, Rafaelsen OJ (1984) Brain and CSF-adrenaline in isocarboxazide and zimeldine treated rats. In: Usdin E (ed) Catecholamines 3. Liss, New York

Hökfelt T, Fuxe K, Goldstein M, Johansson O (1974) Immunohistochemical evidence for the existence of adrenaline neurons in the rat brain. Brain Res 66:235–251

Hökfelt T, Johansson O, Ljungdahl A, Lundberg JM, Schultzberg M (1980) Peptidergic neurons. Nature 284:515

Rafaelsen OJ (1980) Biology of manic-melancholic disorders. Med J Aust 1:637

Roth KA, McIntire SL, Lorenz RG, Barchas JD (1982) Hypothalamic catecholamine changes under acute stress occur independently of nicotinic stimulation. Neurosci Lett 28:47–50

Biochemical and Neuroendocrine Studies in Schizophrenics: Attempts to Characterize the Illness Biochemically

M. ACKENHEIL, M. ALBUS, B. BONDY, F. MÜLLER-SPAHN, U. MÜNCH, and D. NABER

Introduction

The last three decades of research in biochemical psychiatry have been character-ized by many efforts to prove the dopamine (DA) hypothesis of schizophrenia in connection with the effect and supposed mechanism of action of neuroleptic drugs (NLs) (Matthysee 1973). This hypothesis was sustained by clinical results, both by the exacerbating effect of DA agonists such as L-dopa and amphetamine and by the antipsychotic effect of neuroleptic drugs. for (review see Ackenheil et al. 1980). However, these clinical effects are restricted to certain symptoms of schizophrenia and nosological subcategories and, in general, only productive symptoms and the paranoid hallucinatoric type of schizophrenia are involved, whereas minus symp-toms and the hebephrenic type or the schizophrenia simplex type interact with these substances to a lower extent. Biochemical studies more or less failed to prove this hypothesis. The progress of neuroendocrinological research strategies and of know-ledge of receptor measurements promises a better understanding of the patho-physiology of schizophrenia.

Neuroendocrine Studies

Disturbances of the dopaminergic system in the hypothalamic pituitary axes can be evaluated by the apomorphine test, since apomorphine (0.5 mg s.c.), depending on the sensitivity of DA receptors (for review see Meltzer et al. 1981), stimulates the growth hormone (GH) secretion. As reported by other research groups (Rotrosen et al. 1977), acute untreated schizophrenic patients in the mean showed higher GH stimulation values than controls (Fig. 1), although there were great individual vari-ations ranging from blunted GH response to high values. There was no relation be-tween maximum GH response values and ICD diagnostic criteria such as paranoid hallucinatoric schizophrenia and hebephrenia, nor between GH response values and schizophrenic symptoms rated with the BPRS (Brief Psychiatric Rating Scale) and AMDP (Arbeitsgemeinschaft für Methodik und Dokumentation in der Psy-chiatrie) systems. Treatment of these patients with NLs suppressed the GH response in all of them. Withdrawal of NL treatment for 6 days resulted in a slight increase in GH response. There was, however, no correlation between the initial GH stimu-lation and clinical outcome (Table 1).

Long-term treatment with NLs leads to adaptation phenomena of the DA system. Therefore, investigation of the effects of withdrawal of NLs is appropriate to clarify the role of the DA system in the pathophysiology of schizophrenia. For this purpose the GH secretion of 15 patients was tested by apomorphine stimulation during NL treatment (day 0) and after 12 (day 12) and 30 (day 30) days' withdrawal. In these patients prolactin (PRL) secretion, which is normally elevated after acute NL treatment, was in a normal range at day 0. However, PRL secretion diminished from 18.5 ± 9.3 ng/ml to 3.3 ± 7.0 ng/ml at day 12 and to 3.5 ± 3.2 ng/ml at day 30. The application of the DA agonist apomorphine led to a further decrease in PRL secretion from 11.1 to 8.5 ng/ml at day 0; at day 12 from 3.3 to 2.4 ng/ml; at day 30 from 3.5 to 3.2 ng/ml. There was a significant increase of GH secretion after apomorphine stimulation in every period of the investigation (Fig. 1). Significantly higher GH maxima were measured on day 12 (15.9 ± 12.2 ng/ml) and day 30

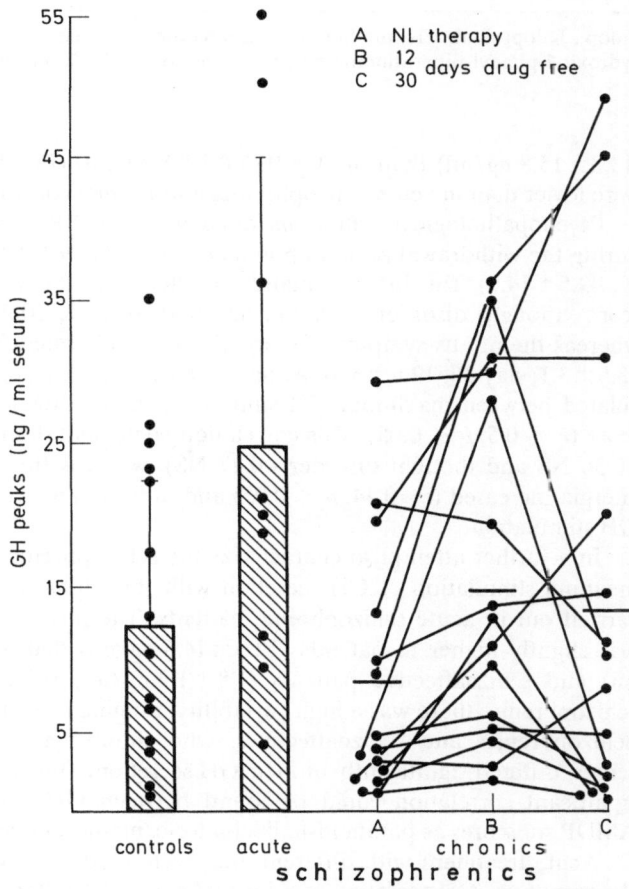

Fig. 1. Apomorphine stimulation (0.5 mg s.c.) of GH secretion in schizophrenic patients (GH maxima) ($n = 10$), controls (age- and sex-matched ($n = 14$), and chronic schizophrenics ($n = 15$). *NL*, neuroleptic

Table 1. GH response to apomorphine

Pat.	Sex	Age	Diagnosis	GH (ng/ml)	Treatment	Duration of treat. (days)	Therap. resp.
1	M	49	295.3	34.9	FK Halop.	48	–
2	F	28	295.3	17.0	Halop.	90	+ / –
3	M	25	295.1	16.2	FK, EMD Semap	90	–
4	M	20	295.3	69.1	Halop.	55	–
5	M	32	295.4	3.3	EMD	30	+
6	F	25	295.3	10.0	Semap	55	–
7	M	29	295.3	5.3	EMD	40	+
8	M	28	295.0	48.6	Halop.	60	+
9	M	31	295.3	20.2	Halop.	8	–
10	M	22	295.1	0.6	Halop.	34	–

Halop., haloperidol; FK, methionine-eukephaline analogue; Semap, penfluridol; EMD, 2-(4-hydroxy-4-phenyl-piperidinomethyl)-6,7-dimethoxy-1,2,3,4,-tetrahydronaphthalinon-HCl

(13.3 ± 15.8 ng/ml) than on day 0 (7.8 ± 8.8 ng/ml) (Fig. 1). However, these values were lower than in acute schizophrenics and normal controls.

Psychopathological symptoms rated with the BPRS system increased slightly during the withdrawal period up to day 12 (day 0, 49 ± 9.7; day 12, 55.2 ± 13.3; day 30, 49.5 ± 14.6). This increase mainly resulted from the worsening of the BPRS sub-score "thought disorders" (day 0, 8.6 ± 4.0; day 12, 10.2 ± 5.6; day 30, 9.0 ± 5.7), whereas the minus symptom "anergia" was ameliorated (day 0, 14.8 ± 3.0; day 12, 13.5 ± 3.1; day 30, 12.6 ± 3.3). At day 12 a significantly negative correlation was calculated between maximum GH stimulation and BPRS total and thought disorder score ($r = -0.5$, $p < 0.05$). This correlation coefficient diminished until day 30 (total -0.36 NS and thought disorder -0.27 NS), whereas the correlation coefficient for anergia increased ($r = 0.44$, $p < 0.05$) and indicated a significant relationship to the GH stimulation.

In a further attempt to characterize the schizophrenic illness, the clonidine test, meaning stimulation of GH secretion with this alpha-2 agonist (0.15 mg i.v.), was carried out in acute schizophrenic patients. The mean maximum GH stimulation was slightly higher in patients (12.0 ± 14.6 ng/ml) than in controls (10.3 ± 8.5 ng/ml) and schizoaffective patients (1.8 ± 1.2 ng/ml) (Fig. 2). Within the group of schizophrenics there was a high variability, ranging from 1.2 to more than 45 ng/ml. Schizophrenics and schizoaffectives, who showed no differences in the AMDP scores, differed significantly in their GH secretion. Thus it is understandable that no significant correlation could be found between GH stimulation levels and such AMDP subscores as paranoid-hallucinatoric, manic, and apathic syndromes.

Acute treatment with different antipsychotic drugs did not significantly change the maximum GH secretion ($n = 17$; GH max, 10.1 ± 10.4 ng/ml). Actually, six of 17 patients showed a GH peak even before the application of clonidine, which probably indicates a higher lability of the alpha-adrenergic system. Chronic schizophrenic patients, who had been treated with NLs for more than 5 years, showed a diminished GH secretion response to clonidine stimulation (0.15 mg i.v.; max, 4.8 ± 6.0 ng/ml) compared to acute schizophrenics and matched controls (GH max, 9.7 ± 9.0 ng/ml) (Fig. 3). Discontinuation of the NL treatment for 5 and 10 days did not induce changes in GH stimulation levels, although most of the patients reacted to the discontinuation with changes in their symptomatology, in both directions. They either improved, mainly due to minus symptomatology, or they worsened due to alterations in productive symptoms.

At the same time as the clonidine results the norepinephrine (NE) secretion in blood was measured. In comparison with controls, acute untreated schizophrenic patients showed increased NE levels during rest and during different stress situations which indicate higher activation, the so-called overarousal. This was also noticed when psychophysiological variables such as electromyogram and heart rate were measured. NL treatment induced an elevation of NE secretion, which was most pronounced after long-term treatment (Fig. 4). Similarly, the levels of the main NE metabolite, 3-methoxy-4-hydroxyphenylglycol (MHPG), in the cerebrospinal fluid were diminished after treatment with 600 mg clozapine per day (Ackenheil et al. 1974). This elevation of NE secretion, resulting most probably from the alpha-adrenolytic effect of NLs, decreases after 15 days' discontinuation of the treatment. No direct significant correlations could be found, either to the symptomatology or to the response to NL treatment. However, in order to evaluate the predictive value of NE levels after long-term neuroleptic therapy, we measured PRL and cortisol in the 12 patients with the highest and in the 12 with the lowest NE levels out of a sample

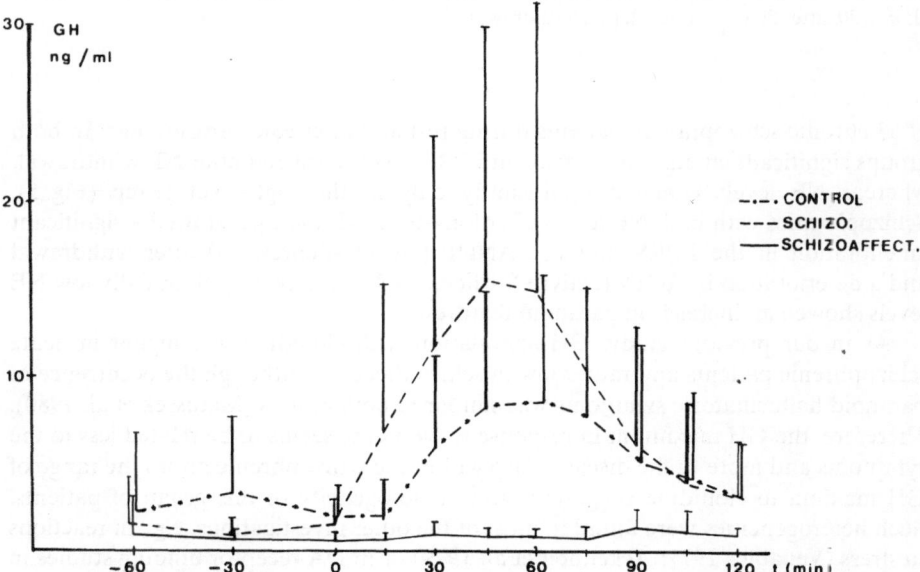

Fig. 2. GH stimulation with clonidine (0.15 mg i.v.). Acute schizophrenics (13) and schizoaffectives (9) compared to age- and sex-matched controls (23). $p < 0.05$; Mann-Whitney U-test. Schizoaffectives vs controls; schizoaffectives vs schizophrenics

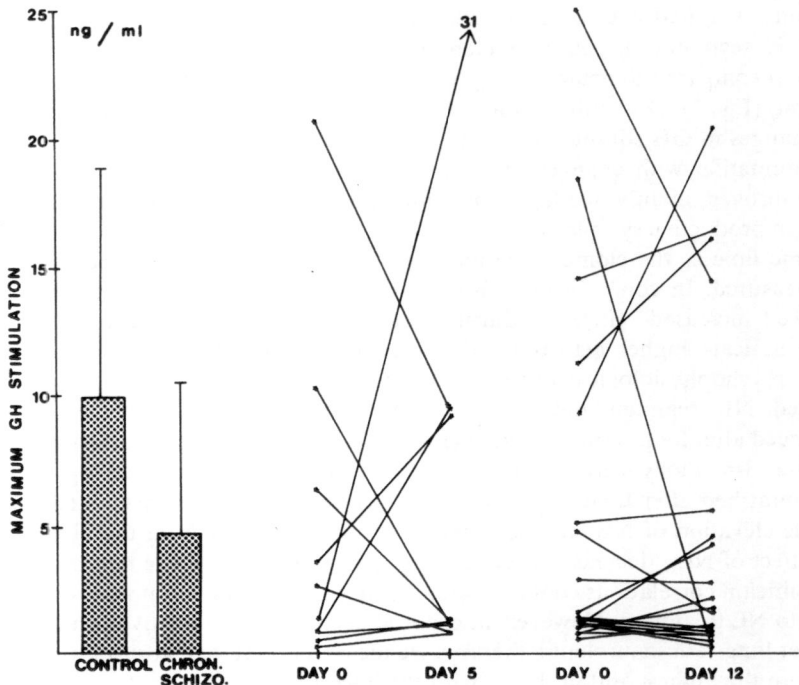

Fig. 3. GH stimulation with clonidine (0.15 mg i.v.). Chronic schizophrenics (*n*=28; age 44.7±10); controls (*n*=23; age 42.7±14). Total group during long-term neuroleptic treatment. *Day 0*, during longterm neuroleptic treatment; *day 5*, after 5 days' neuroleptic withdrawal; *day 30*, after 30 days' neuroleptic withdrawal

of 53 chronic schizophrenic patients during rest and after cold pressure test. In both groups significant changes in cortisol and PRL levels occurred after NL withdrawal, whereas NE levels changed significantly only in the high-level group (Fig. 5). Schizophrenics with high NE levels after long-term NL therapy showed a significant amelioration in the BPRS subscore ANDP (anxiety-depression) after withdrawal and a deterioration in ACTV (activity), whereas schizophrenics with initially low NE levels showed an increase in paranoid features.

As in our previous study, GH stimulation with clonidine was higher in acute schizophrenic patients and rather low in schizoaffectives, although the occurrence of paranoid hallucinatoric symptoms was similar in both groups (Matussek et al. 1980). Therefore, the GH maximum in response to clonidine seems to be related less to the symptoms and more to the disease. Also within the schizophrenic group the range of GH maxima to clonidine suggests a great heterogeneity of this group of patients. Such heterogeneities were found in most of the other investigations, e.g., in reactions to stress (Venables 1977; Ackenheil et al. 1979) or in DA receptor binding studies in postmortem brains (Cross et al. 1978), as well as in GH stimulation studies with apomorphine (Ackenheil 1981; Pandey et al. 1977; Rotrosen et al. 1977). In this study we were not able to divide the schizophrenic patients according to the GH

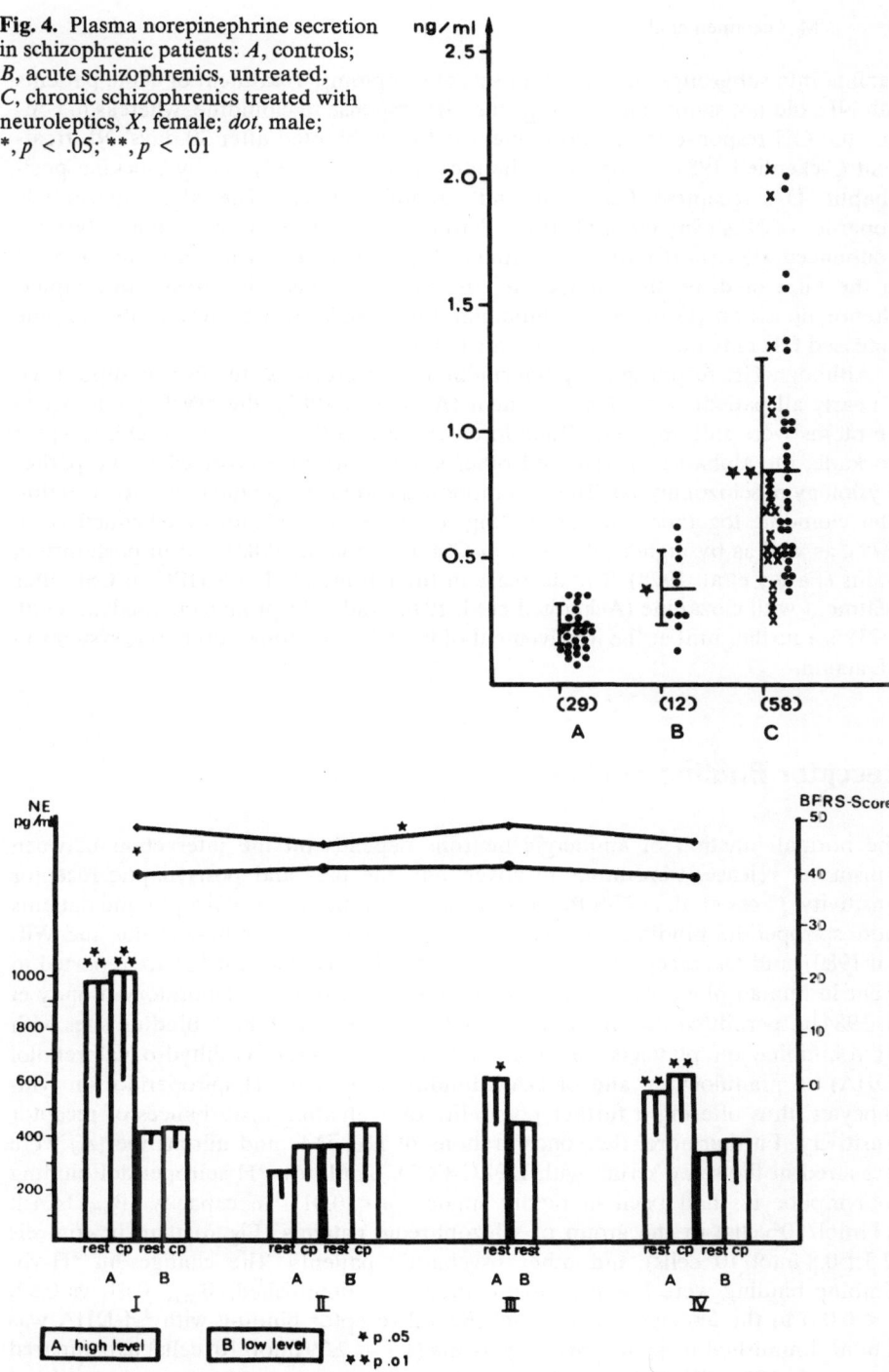

Fig. 4. Plasma norepinephrine secretion in schizophrenic patients: *A*, controls; *B*, acute schizophrenics, untreated; *C*, chronic schizophrenics treated with neuroleptics, *X*, female; *dot*, male; *, *p* < .05; **, *p* < .01

Fig. 5. Plasma norepinephrine (*NE*) and BPRS score during and after withdrawal of neuroleptics. Day *I*, under neuroleptics; *II*, wash out; *III*, CGP-Sulp; *IV*, under neuroleptics. Twenty-four chronic schizophrenics, 12 with high and 12 with low NE levels. *A*, high-level; *B*, low-level; cp, cold pressor test. *p < 0.05; **p < 0.01; ***p < 0.001

maxima into subgroups with different sets of symptoms. Treatment of acute patients with NLs did not significantly change the GH response to clonidine, whereas in contrast the GH response to apomorphine was totally blunted after 30 days' NL treatment (Ackenheil 1981), supporting the idea that NLs mainly act by blocking postsynaptic DA receptors (Carlsson and Lindquist 1963). The alpha-adrenolytic properties of NLs (Snyder et al. 1978; Peroutka et al. 1977) do not seem to be very pronounced after short-term NL treatment (up to 4 weeks). They probably depend on the kind of drug, the dosage and the plasma levels. The presynaptic alpha-adrenolytic action (Gross and Schülmann 1980) could also be responsible for the increased GH maxima occurring in some patients.

Although GH response to apomorphine in our previous studies was suppressed in nearly all patients after NL treatment (Ackenheil 1981), the psychopathological symptoms were still apparent. Therefore, we assume that besides the DA receptor blockade, the alpha-adrenergic and other systems are also involved in the pathophysiology of schizophrenia. This assumption is supported by the high GH secretion after clonidine together with the findings of stress investigations (Ackenheil et al. 1979), as well as by higher NE levels in CSF (Lake et al. 1980) and in postmortem brains (Farley et al. 1978). The decrease in the NE metabolite MHPG in CSF after treatment with clozapine (Ackenheil et al. 1974) and chlorpromazine (Sedvall et al. 1975) is a further hint at the involvement of the NE and alpha-adrenergic systems in NL action.

Receptor Binding Studies

The normal function of aminergic neurons depends on the interaction between transmitter release, transmitter turnover, and the pre- and postsynaptic receptor sensitivity. Cross et al. (1978) found in postmortem brains of schizophrenic patients more spiroperidol binding sites. Alpha 2 (Kafka et al. 1979), beta-2 (Dulis and Wilson 1980), and DA receptors (Le Fur et al. 1982; Bondy et al. 1984a) are reported to occur in human blood cells. A new method, developed in our laboratory (Bondy et al. 1984b), permitted the simultaneous determination of alpha-2 binding sites with ^{3}H-yohimbine on platelets, of beta-2 binding sites with ^{3}H-dihydroxyalprenolol (DHA) on granulocytes, and of DA-2 binding sites with ^{3}H-spiroperidol on lymphocytes, thus offering a further possibility of evaluating disturbances of receptor sensitivity. Furthermore, the concentrations of NE, DA, and adrenaline (A) were measured in the same serum with HPLC-ECD. Until now ^{3}H-spiroperidol binding on lymphocytes had been markedly higher ($p < 0.01$) in capacity (B_{max} 14.4 ± 9.3 fmol/10^6 cells) in the group of schizophrenic patients (Fig. 6) than in controls (2.5 ± 0.8 fmol/10^6 cells) and other psychiatric patients. The changes in ^{3}H-yohimbine binding were less prominent: they were diminished (B_{max} 0.41, vs 0.52, $p < 0.05$) to the 5% significance level. Beta-2 receptor binding with ^{3}H-DHA was similar diminished in schizophrenic patients (1.7 vs 2.79 fmol/10^6 cells) as compared to controls. The affinity for all three binding sites remained almost unchanged between the different patients and control groups. The concentrations of NE, DA, and A showed great interindividual variations, ranging from normal values up to threefold increase. Interestingly, in schizophrenic patients not only high NE values were found (493 ± 349 pg/ml, control 210 ± 40 pg/ml), but also high concentrations of DA in serum (50 ± 47 pg/ml, control 18 ± 9 pg/ml) (Fig. 7). No relation to psychopathological symptoms and nosological subgroups (ICD: 295.1; 295.3; 295.7) has been established until now; probably the small number of patients accounts for this lack. Although according to knowledge about the regulation of synpases a

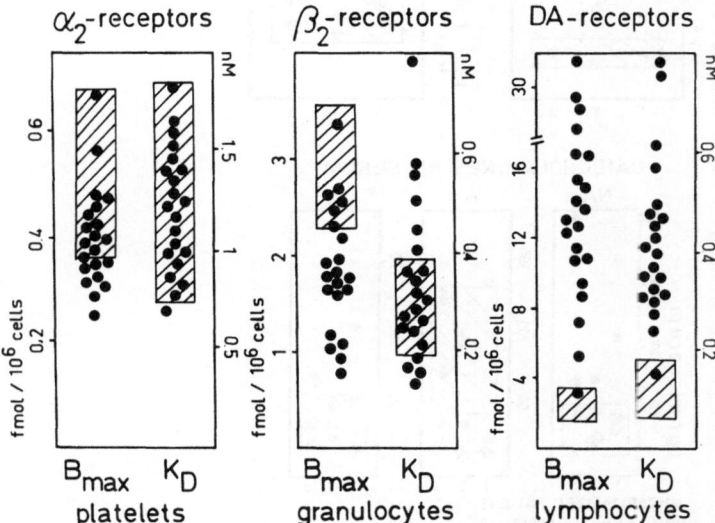

Fig. 6. Alpha-2, beta-2, and dopamine receptor binding on human blood cells in controls (*hatched* area gives range of $\bar{x} \pm$ SD) and acute schizophrenic patients (*dots*)

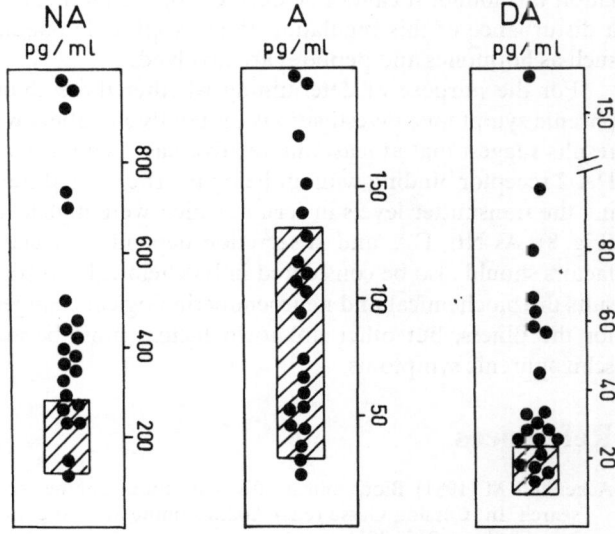

Fig. 7. Plasma norepinephrine (*NA*), adrenaline (A), and dopamine (*DA*) in acute schizophrenic patients (*dots*) and controls (*hatched* area)

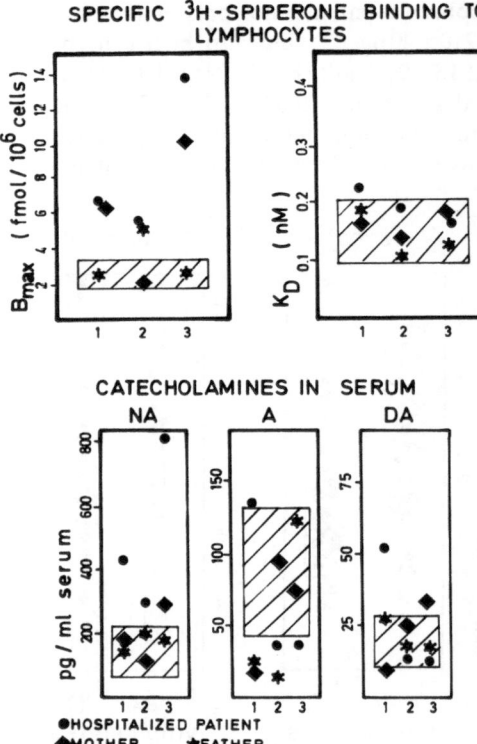

Fig. 8. Alpha-2, beta-2, and dopamine (*DA*) binding on blood cells. Norepinephrine (NA), adrenaline (*A*), and DA in serum of three schizophrenics and their families

correlation should exist between transmitter release and receptors, no such correlation was found. It cannot be decided for the moment whether the reason for this is a disturbance of this regulation in schizophrenic patients or whether other factors such as hormones and peptides are involved.

For the purpose of determining whether these changes are specific for schizophrenic symptoms, investigations on family members were carried out. Preliminary results suggest that at least one relative has a similar pattern of alpha-2, beta-2, and DA-2 receptor binding without being ill. The main differences were seen by measuring the transmitter levels in serum, which were higher in the schizophrenic patients (Fig. 8). As NE, DA, and A secretion depend on social factors such as stress, such factors should also be considered in biochemical investigations. According to our results the biochemical and neuroendocrinological changes measured can be a marker for the illness, but other unknown factors must be added for the appearance of schizophrenic symptoms.

References

Ackenheil M (1981) Biochemical effects of apomorphine: contribution to schizophrenia research. In: Corsini, Gessa (eds) Apomorphine and other dopaminomimetics, vol 2. Raven, New York, pp 215–255

Ackenheil M, Beckmann H, Greil W, Hoffmann G, Markianos E, Raese J (1974) Antipsychotic efficacy of clozapine in correlation to changes in catecholamine metabolism in man. In:

Forrest I, Carr CJ, Usdin E (eds) Phenothiazines and structurally related drugs. Raven, New York, pp 647–657

Ackenheil M, Albus M, Müller F, Müller T, Welter D, Zander K, Engel R (1979) Catecholamine response to short-time stress in schizophrenic and depressive patients. In: Usdin E, Kopin IJ, Barchas J (eds) Catecholamines: basic and clinical frontiers, vol 2. Pergamon, New York

Ackenheil M, Hippius H, Matussek N (1980) Neuere Entwicklungen und Ansätze der biochemischen Schizophrenieforschung. In: Huber G (ed) Stand und Entwicklungstendenzen der Forschung. Schattauer, Stuttgart

Bondy B, Ackenheil M, Elbers R, Fröhler M (1984a) Binding of ^{3}H-spiperone to human lymphocytes: a biological marker in psychiatry? Psychiatry Res (in press)

Bondy B, Ackenheil M, Birzle W, Elbers R, Fröhler M (1984b) Catecholamines and their receptors in blood: evidence for alterations in schizophrenia. Biol Psychiatry (in press)

Carlsson A, Lindquist M (1963) Effect of chlorpromazine and haloperidol on formation of 3-methoxytyramine and normetanephrine in mouse brain. Acta Pharmacol Toxicol 20:140–144

Cross AJ, Crow TJ, Longden A, Owen F, Poulter M, Riley GJ (1978) Evidence for increased dopamine receptor sensitivity in post mortem brains from patients with schizophrenia. J Physiol (Lond) 280:37

Dulis BH, Wilson IB (1980) The β-adrenergic receptor of live human polymorphonuclear leukocytes. J Biol Chem 255 (3):1043–1048

Farley IJ, Price KS, McCullogh E, Deck JHN, Hordynski W, Hornykiewicz O (1978) Norepinephrine in chronic paranoid schizophrenia: above normal levels in limbic forebrain. Science 200:456–458

Groß G, Schümann HJ (1980) Enhancement of noradrenalin release from rat cerebral cortex by neuroleptic drugs. Naunyn-Schmiedebergs Arch Pharmacol 315:103–109

Kafka MS, Van Kammen D, Bunney WMD (1979) Reduced cyclic AMP production in the blood platelets from schizophrenic patients. Am J Psychiatry 136:5

Lake RC, Sternberg DE, Van Kammen DP, Ballenger JC, Ziegler MG, Post RM, Kopin IJ, Bunney WE (1980) Schizophrenia: elevated cerebrospinal fluid norepinephrine. Science 207:331–333

Le Fur G, Zorifian E, Phan T, Cuhe H, Flamier A, Bouchami J, Bourgevin MC, Lov H, Gerard A, Uzan A (1982) [^{3}H]-Spiroperidol binding on lymphocytes: changes in two different groups of schizophrenic patients and effect of neuroleptic treatment. Life Sci 32:249–255

Matthysee S (1973) Antipsychotic drug action: a cue to the neuropathology of schizophrenia? Fed Proc 32:200–205

Matussek N, Ackenheil M, Hippius H, Müller F, Schröder T, Schultes H, Wasilewski B (1980) Effect of clonidine on growth hormone release in psychiatric patients and controls. Psychiatry Res 2:25–36

Meltzer H, Busch B, Vang V (1981) Hormones, dopamine receptors and schizophrenia. Psychoneuroendocrinology 6 (1):17–36

Pandey GN, Garner DL, Tamminga C, Ericksen S, Ali SI, Davis JM (1977) Postsynaptic supersensitivity in schizophrenia. Am J Psychiatry 134:518

Peroutka SJ, Prichard DC, Greenberg DA, Snyder SH (1977) Neuroleptic drug interactions with norepinephrine alpha receptor binding sites in rat brain. Neuropharmacology 6:549–556

Rotrosen J, Angrist BM, Gershon S, Sachar EJ, Halpern FS (1977) Neuroendocrine assessment of dopaminergic activity in schizophrenia. In: Costa, Gessa (eds) Advances in biochemical psychopharmacol, vol 16. Raven, New York

Sedvall G, Alfredsson G, Bjerkenstedt L, Eneroth P, Fyrö B, Härnryd CG, Swahn FA, Wode-Helogdt B (1975) Selective effects of psychoactive drugs on levels of monoamine metabolites and prolactin in cerebrospinal fluid of psychiatric patients. Proceedings of the 6th international congress of pharmacology, 1973, Helsinki

Snyder SH, Prichard DC, Greenberg DA (1978) Neurotransmitter receptor binding in the brain. In: Lipton, Di Mascio, Killam (eds) Psychopharmacology: a generation of progress. Raven, New York, pp 361–370

Venables PH (1977) The psychophysiology of schizophrenics and children at risk for schizophrenia: controverses and developments. Schizophr Bull 3 (1):28–48

Decreased Spinal Fluid Monoamine Metabolites and Norepinephrine in Schizophrenic Patients with Brain Atrophy

D. P. Van Kammen, Lee S. Mann, Mika Scheinin, Philip T. Ninan, Welmoet B. Van Kammen, and Markku Linnoila

Introduction

Spinal fluid studies in schizophrenia have not led to the discovery of consistent differences from normals and other comparison groups (Berger et al. 1980; Post et al. 1975). The frequently observed increased variance in the CSF of schizophrenic patients has been explained by the heterogeneous nature of this illness. Attempts to subgroup schizophrenic patients on the basis of CSF has led to associations between reduced homovanillic acid (HVA) accumulations after probenecid and patients with a poor prognosis (Bowers 1973), decreased HVA and remitted acute schizophrenics (Post et al. 1975), and elevated norepinephrine (NE) and paranoid patients (Lake et al. 1980). Post et al. (1975) found a correlation between 5-hydroxyindoleacetic acid (5-HIAA) concentrations and Schneiderian first-rank symptoms in acutely ill schizophrenics, and we recently found low 5-HIAA in a small group of schizophrenics with a history of successful or serious suicide attempts (Ninan et al. 1984).

Controlled computerized tomographic (CT) studies have found that structural brain abnormalities can be observed in some schizophrenic patients (Rieder et al. 1979; Weinberger et al. 1979a, b; Golden et al. 1980; Andreasen et al. 1982). These abnormalities seem to be unrelated to age, chronic drug treatment, or duration of illness. Some studies have shown very high ventricle-brain ratios (VBRs) in adolescent schizophrenic and schizophreniform patients, while some older chronic schizophrenic patients have very small ventricles (Weinberger et al. 1982; Golden et al. 1980). Brain abnormalities therefore strongly suggested the notion that schizophrenia is not a single disease entity but a group of disorders that influence brain function. Other characteristics of patients with brain abnormalities were explored, and it was reported that patients with enlarged cerebral ventricles tended to perform poorly on neuropsychological tests (Golden et al. 1980; Johnstone et al. 1976), had poorer premorbid adjustment (Weinberger et al. 1980), and displayed more negative symptoms (Andreasen et al. 1982). They also seemed to respond less well to neuroleptic drugs.

In this paper we report on the association between brain atrophy and CSF HVA, dihydroxyphenylacetic acid (DOPAC), 5-HIAA, and NE in drug-free schizophrenic patients. Our study is a further attempt to separate patients with brain abnormalities into biologically and clinically meaningful subgroups and to explain the increased variance of CSF in schizophrenia (Linnoila et al. 1983).

Methods

All 56 patients in this study were voluntarily admitted to the schizophrenia unit of the Clinical Center at the National Institutes of Health after having given informed consent. Each subject met the Research Diagnostic Criteria (RDC) and the DSM-III requirements for the diagnosis of schizophrenia.

Patients were off medication for a minimum of 14 days (mean 32.4 days) before the lumbar puncture (LP) was performed and were kept on a low-monoamine, alcohol-free, caffeine-restricted diet. All LPs were performed between 8:30 and 9:30 in the morning. Spinal fluid was collected from lumbar space between the third and fourth vertebrae with patients in lateral decubitus position after fasting and bed rest since the previous night. Twelve milliliters of CSF was collected and placed on ice at bedside, to be subsequently aliquoted and frozen at $-60\,^{\circ}\text{C}$ until assayed. The CSF levels were measured using high-performance liquid chromatography with electrochemical detection in two 1-ml aliquots of the CSF pool. Details of these assays have been published elsewhere (Petrucelli et al. 1982; Linnoila et al. 1983).

The CT scans, all without contrast, yielded 12 radiographic images ($160' \times 160'$ matrix) of different brain levels and were performed on the EMI-1010 ($n = 40$) and GE-8800 ($n = 16$) at the Clinical Center. Patients with a history of alcoholism, head trauma, or other possible causes of brain atrophy were excluded from this study. Analysis of the CT scans included quantitative measurements of ventricle size, i.e., VBR (Weinberger et al. 1979 a), and measures of atrophy of cortical fissures and sulci (Rieder et al. 1979). For the purpose of this study, all patients ($n = 5$) whose VBR readings were 2 SDs above the mean of the VBR measurements of a neurological control population (8.4%) or who had evidence of abnormal sulci ($n = 10$) were placed in the abnormal CT scan group ($n = 13$). All others were considered to have normal CT scans. Two patients had elevated VBRs as well as abnormal sulci. When we started our study in 1978, we decided on the cutoff point for enlarged ventricle size based on the findings of Weinberger et al. (1979 a). Since then other cutoff points have been used, ranging from 5.8% to 10% (Andreasen et al. 1982). Inspection of our correlational graphs restrospectively indeed suggests that 8.3% is a reasonable cutoff point for normal VBRs.

Results

The 5-HIAA ($p = 0.0001$), DOPAC ($p = 0.0026$), and HVA ($p = 0.0001$) concentrations in the CSF of patients with brain atrophy were significantly lower than the CSF values of the patients with normal brain scans (Table 1). When we looked at the ventricle size and cortical atrophy separately, 5-HIAA correlated inversely with both ventricle size ($r = -0.35$, $p = 0.009$; Fig. 1) and cortical atrophy ($r = -0.38$, $p = 0.006$; Fig. 2). There was also a negative correlation between cortical atrophy and HVA ($r = -0.49$, $p = 0.0003$), NE ($r = -0.29$, $p = 0.04$), and DOPAC ($r = -0.39$, $p = 0.02$; Fig. 3). The mean of 5-HIAA ($p = 0.01$) was significantly lower in patients with high VBRs, as were the means of 5-HIAA ($p = 0.0001$), HVA ($p = 0.0002$; Fig. 4), NE ($p = 0.0038$), and DOPAC ($p = 0.01$) in patients with cortical atrophy.

Table 1. Statistical relationship between CSF monoamine metabolites and brain abnormalities

	VBR		Cortical atrophy		VBR			
	r	p	r	p	Normal		High		p
HVA pmol/ml (N)	− 0.23 (56)	0.09	− 0.49 (50)	0.0003	177 (51)	± 66.6	127 (5)	± 46.4	NS
5-HIAA pmol/ml (N)	− 0.35 (56)	0.009	− 0.38 (50)	0.006	103 (51)	± 39.5	69 (5)	± 20.8	0.01
DOPAC pmol/ml (N)	− 0.05 (37)	NS	− 0.39 (33)	0.02	2.4 (33)	± 1.03	1.8 (4)	± 0.50	NS
NE pmol/ml (N)	− 0.04 (54)	NS	− 0.29 (48)	0.04	0.69 (49)	± 0.289	0.64 (5)	± 0.348	NS
MHPG pmol/ml (N)	− 0.02 (54)	NS	− 0.18 (48)	NS	40.0 (49)	± 12.68	36.8 (5)	± 3.74	NS

VBR, ventricle-brain ratios; HVA, homovanillic acid; 5-HIAA, 5-hydroxyindoleacetic acid; DOPAC, NE, norepinephrine; MHPG, 3-methoxy-4-hydroxyphenylglycol

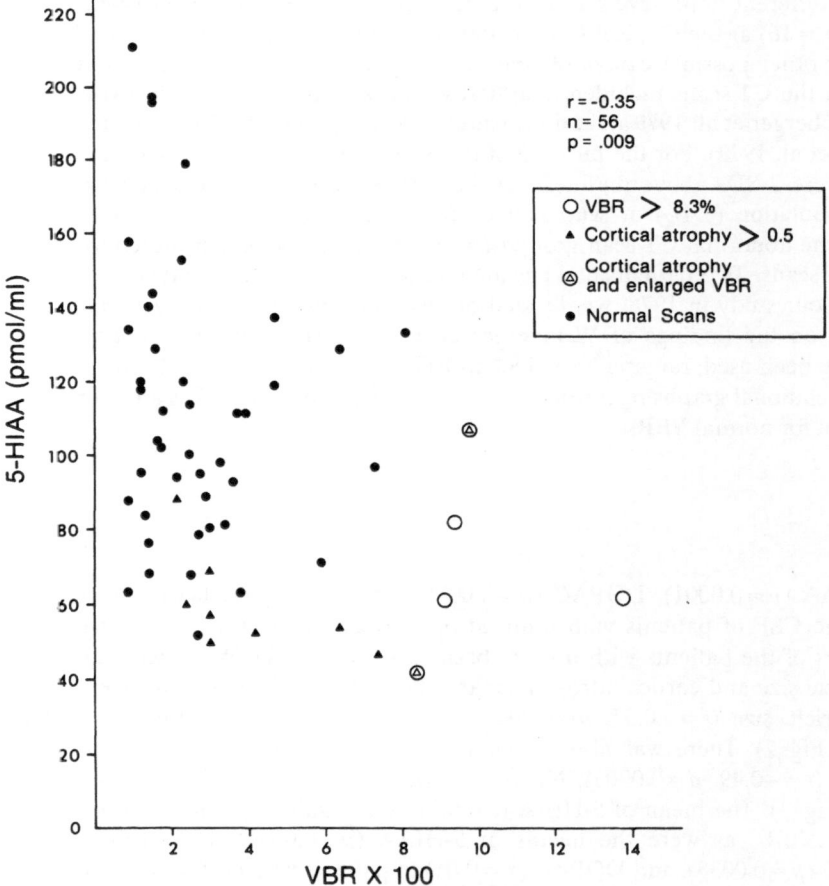

$r = -0.35$
$n = 56$
$p = .009$

○ VBR > 8.3%
▲ Cortical atrophy > 0.5
⊛ Cortical atrophy and enlarged VBR
● Normal Scans

Fig. 1. CSF 5-HIAA correlated significantly with VBR

Cortical atrophy			Brain atrophy		
Absent	Present	p	Normal	Abnormal	p
188 ±64.3 (40)	104 ±36.8 (10)	0.0002	192 ±60.9 (43)	109 ±38.8 (13)	0.0001
109 ±39.0 (40)	62 ±18.0 (10)	0.0001	112 ±37.3 (43)	63 ±16.5 (13)	0.0001
2.6 ± 1.02 (24)	1.7 ± 0.60 (9)	0.01	2.6 ± 1.03 (26)	1.7 ± 0.59 (11)	0.0026
0.74± 0.260 (38)	0.48± 0.175 (10)	0.0038	0.72± 0.294 (41)	0.55± 0.253 (13)	0.06
40.5 ±13.35 (39)	35.8 ± 5.53 (9)	NS	40.6 ±13.41 (42)	36.6 ± 5.12 (12)	NS

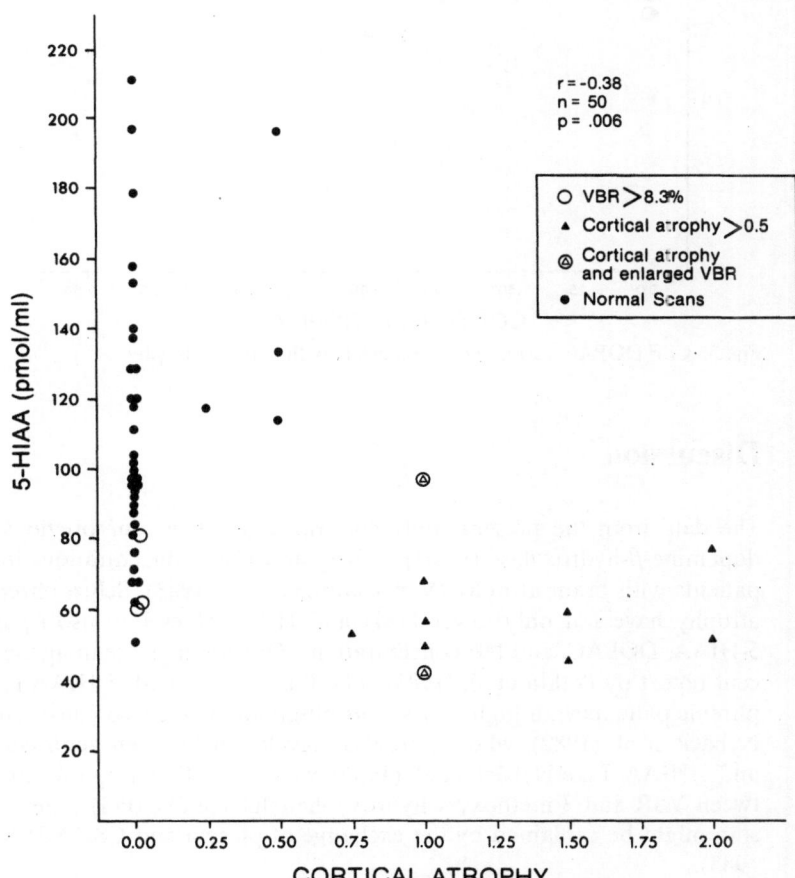

Fig. 2. CSF 5-HIAA correlated significantly with cortical atrophy

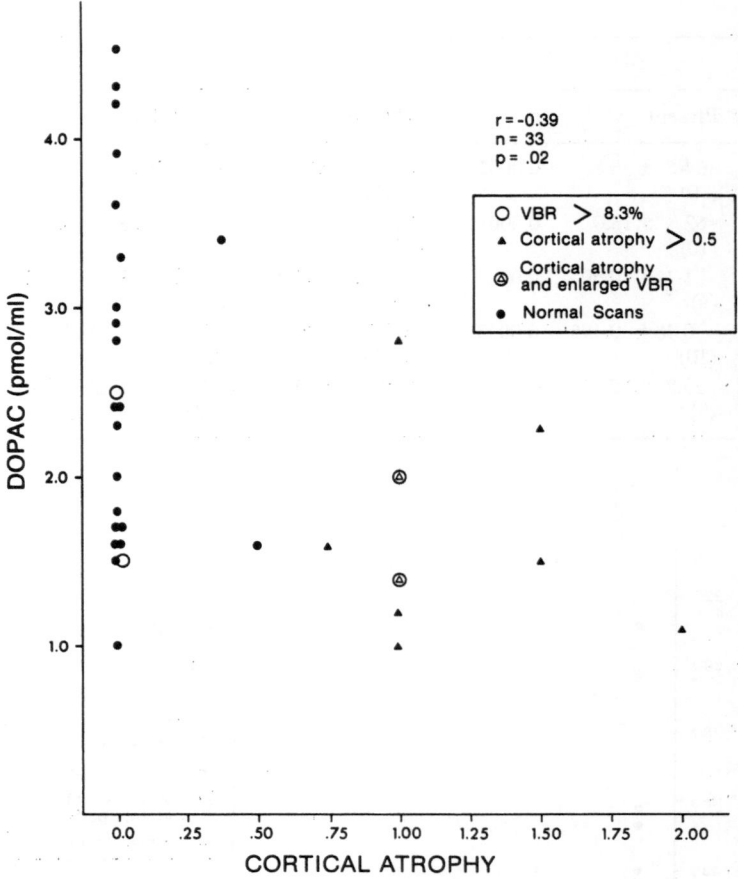

Fig. 3. CSF DOPAC correlated significantly with cortical atrophy

Discussion

The data from the present study confirm and extend our previous report of low dopamine-β-hydroxylase (DBH) activity and HVA concentrations in schizophrenic patients with brain atrophy (Van Kammen et al. 1983). Schizophrenics with brain atrophy have not only lower DBH and HVA values but also significantly lower 5-HIAA, DOPAC, and NE concentrations. Our findings are in agreement with a recent report by Potkin et al. (1983), who found lower CSF 5-HIAA values in schizophrenic patients with high VBRs. Our observations are also consistent with those of Nybåck et al. (1982), who reported a correlation between ventricle size and HVA and 5-HIAA. Like Nybåck et al. (1982), we did not find a significant correlation between VBR and 3-methoxy-4-hydroxyphenylglycol (MHPG). This lack of relationship might be explained by the exchange of plasma and CSF MHPG (Kopin et al. 1983).

Fig. 4. CSF HVA was significantly decreased in patients with brain atrophy

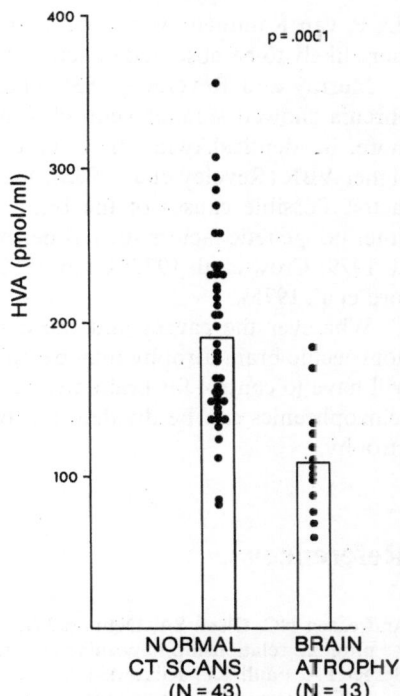

The decreased CSF monoamine values suggest a global monoamine disturbance. Brain monoamine systems interact with each other. Therefore, the question which system is primarily affected in schizophrenia remains unanswered. Usually, it is presumed that the striatum provides most of the spinal fluid HVA and DOPAC. However, Elsworth et al. (1983) reported in monkeys that CSF HVA correlated significantly with dorsal and retro-orbital frontal cortex HVA but not with striatal HVA content. Therefore, low CSF DOPAC and HVA may very well reflect a frontal cortex disturbance, but could include disturbances in the corticostriatal pathways as well. Kim et al. (1980) reported decreased CSF glutamate levels in schizophrenic patients and concluded that the increased dopamine D_2 receptors that are located on the corticostriatal glutamate neurons were responsible for this decrease. Conceivably, cortical cell loss could explain the decreased glutamate levels as well. Cortical atrophy is certainly consistent with the reported decreased frontal blood flow (Ingvar and Franzen 1974; Mathew et al. 1982) and the relatively decreased frontal occipital ratio of metabolic activity with positron emission tomography (PET) scan in schizophrenic patients (Buchsbaum et al. 1982).

It is possible that the decreased CSF values can be explained by dilution: in patients with atrophy CSF constituents are diluted because of larger fluid space. However, spinal fluid concentrations of protein and peptides (M. Linnoila, M. Sheinin, L. Mann, and W. B. van Kammen, May 1982, unpublished observations) were identical for the two groups, which speaks against dilution as a major source for the variance. Not only that, but CSF cortisol was higher in patients with increased VBR

(D. P. van Kammen, May, 1982, unpublished observations). Presumably, dilution is more likely to be observed in acute atrophy.

Murray and Reveley (1983) reported that schizophrenics with familial schizophrenia showed smaller ventricles than those without a genetic loading. Furthermore, in identical twins discordant for schizophrenia the ill twin usually has the higher VBR (Reveley et al. 1982), suggesting that atrophy is not caused by a genetic factor. Possible causes of the brain atrophy could be a previous encephalitis, or other nongenetic factors such as perinatal anoxia or trauma (Stevens 1982; Tyrrell et al. 1979; Crow et al. 1979; Crow 1982; Torrey et al. 1982; Rieder et al. 1975; Handford et al. 1975).

Whatever the causes and the clinical implications of the subtle and probably nonspecific brain atrophy may be, future CSF monoamine studies of schizophrenics will have to control for brain atrophy. At present, our data support the concept that schizophrenics can be divided into two types, i.e., patients with and without brain atrophy.

References

Andreasen NC, Olsen SA, Dennert JW, Smith MR (1982) Ventricular enlargement in schizophrenia: relationship to positive and negative symptoms. Am J Psychiatry 139:297–301

Berger PA, Faull KF, Kilkowski J, Anderson PJ, Kraemer H, Davis KL, Barchas JD (1980) CSF monoamine metabolites in depression and schizophrenia. Am J Psychiatry 137:174–180

Bowers MB (1973) 5-Hydroxyindoleacetic acid (5-HIAA) and homovanillic acid (HVA) following probenecid in acute psychotic patients treated with phenothiazines. Psychopharmacologia 28:309–318

Buchsbaum MS, Ingvar DH, Kessler R, Waters RN, Cappelletti J, van Kammen DP, King AC, Johnson JL, Manning RG, Flynn RW, Mann LS, Bunney WE Jr, Sokoloff L (1982) Cerebral glucography with positron tomography. Arch Gen Psychiatry 39:251–259

Crow T (1982) Biological basis of mental disorders: the case of viral aetiology. In: Namba M, Kaiya H (eds) Psychobiology of schizophrenia. Pergamon, Oxford, pp 249–263

Crow TJ, Ferrier IN, Johnstone EC, Macmillan JF, Owens DGC, Parry RP, Tyrrell DA (1979) Characteristics of patients with schizophrenia or neurological disorder and virus-like agent in cerebrospinal fluid. Lancet 1:842–844

Elsworth JD, Roth RM, Redmond DE (1983) Does HVA concentration in CSF or plasma reflect central dopamine function in primates? Scientific proceedings of the 5th catecholamine conference, June 1983, Göteborg, Sweden. 125 (no. 128)

Golden CJ, Moses JA, Zelazowski R, Graber B, Zatz LM, Horvath TB, Berger PA (1980) Cerebral ventricular size and neuropsychological impairment in young chronic schizophrenics. Arch Gen Psychiatry 37:619–623

Handford HA (1975) Brain hypoxia, minimal brain dysfunction and schizophrenia. Am J Psychiatry 132:192–194

Ingvar DH, Franzen G (1974) Abnormalities in cerebral blood flow distribution in patients with chronic schizophrenia. Acta Psychiatr Scand 50:425–462

Johnstone EC, Crow TL, Frith CD, Husband J, Kreel L (1976) Cerebral ventricular size and cognitive impairment of chronic schizophrenia. Lancet 2:924–925

Kim JS, Kornhuber HH, Schmid-Burk W, Holzheimer B (1980) Low cerebrospinal fluid glutamate in schizophrenic patients and a new hypothesis on schizophrenia. Neurosci Lett 20:379–382

Kopin IJ, Gordon EK, Jimerson DC, Polinsky RJ (1983) Relation between plasma and cerebrospinal fluid levels of 3-methoxy-4-hydroxyphenylglycol. Science 219:73–75

Lake CR, Sternberg DE, Van Kammen DP, Ballenger JC, Ziegler MG, Post RM, Kopin IJ, Bunney WE Jr (1980) Schizophrenia: elevated cerebrospinal fluid norepinephrine. Science 207:331–333

Linnoila M, Ninan PT, Scheinin M, Chang WH, Waters RN, Van Kammen DP (1983) Reliability of norepinephrine and major monoamine metabolite measurements in CSF of schizophrenic patients. Arch Gen Psychiatry 40:1290–1299

Mathew RJ, Duncan GC, Weinman ML, Barr DL (1982) Regional cerebral blood flow in schizophrenia. Arch Gen Psychiatry 39:1121–1124

Murray RM, Reveley AM (1983) Genetics of schizophrenia (reply) Lancet 1:1159–1160

Ninan PT, Van Kammen DP, Scheinin M, Linnoila M, Bunney WE Jr, Goodwin FK (1984) Cerebrospinal fluid 5-HIAA in suicidal schizophrenic patients. Am J Psychiatry 141:566–569

Nybåck H, Wiesel FA, Hindmarsh T, Sedvall G (1982) Cerebroventricular volume and cerebrospinal fluid monoamine metabolites in psychotic patients and in healthy volunteers. Col Inter Neuro-psychopharmacologicum (Abstracts) 2:534

Petruccelli B, Bakris G, Miller T, Korpi ER, Linnoila M (1982) A liquid chromatic assay for 5-hydroxytryptophan serotonin and 5-hydroxyindoleacetic acid in human body fluids. Acta Pharmacol Toxicol 51:421–427

Post RM, Fink E, Carpenter WT, Goodwin FK (1975) Cerebrospinal fluid amine metabolites in acute schizophrenia. Arch Gen Psychiatry 32:1063–1069

Potkin SG, Weinberger DR, Linnoila M, Wyatt RJ (1983) Low CSF 5-hydroxyindoleacetic acid in schizophrenic patients with enlarged cerebral ventricles. Am J Psychiatry 140:21–25

Reveley AM, Reveley MA, Clifford CA, Murray RM (1982) Cerebral ventricular size in twins discordant for schizophrenia. Lancet 1:540–541

Rieder RO, Rosenthal D, Wender P, Blumenthal H (1975) The offspring of schizophrenics. Arch Gen Psychiatry 32:200–211

Rieder RO, Donnelly EF, Herdt JR, Waldman IN (1979) Sulcal prominence in young schizophrenic patients: CT scan findings associated with impairment on neuropsychological tests. Psychiatry Res 18:1–8

Stevens J (1982) Neuropathology of schizophrenia. Arch Gen Psychiatry 39:1131–1139

Torrey EF, Yolken RH, Winfrey CJ (1982) Cytomegalovirus antibody in cerebrospinal fluid of schizophrenic patients detected by enzyme immunoassay. Science 216:892–893

Tyrrell DA, Parry RP, Crow TJ, Johnstone EC, Ferrier IN (1979) Possible virus in schizophrenia and some neurological disorders. Lancet 1:839–841

Van Kammen DP, Mann LS, Sternberg DE, Scheinin M, Ninan PT, Marder SR, Van Kammen WB, Rieder RO, Linnoila M (1983) Dopamine-β-hydroxylase activity and homovanillic acid in spinal fluid of schizophrenics with brain atrophy. Science 220:974–977

Weinberger DR, Torrey EF, Neophytides AN, Wyatt RJ (1979a) Lateral cerebral ventricular enlargement in chronic schizophrenia. Arch Gen Psychiatry 36:735–739

Weinberger DR, Torrey EF, Neophytides AN, Wyatt RJ (1979b) Structural abnormalities in cerebral cortex of schizophrenic patients. Arch Gen Psychiatry 36:935–939

Weinberger DR, Cannon-Spoor E, Potkin SG, Wyatt RJ (1980) Poor premorbid adjustment and CT scan abnormalities in chronic schizophrenia. Am J Psychiatry 137:1410–1413

Weinberger DR, DeLisi LE, Neophytides AN, Wyatt RJ (1981) Familial aspects of CT scan abnormalities in chronic schizophrenic patients. Psychiatric Res 4:65–71

Weinberger DR, DeLisi LE, Perman GP, Targum S, Wyatt RJ (1982) Computed tomography in schizophreniform disorder and other acute psychiatric disorders. Arch Gen Psychiatry 39:778–783

The Use of Platelet Monoamine Oxidase in Multifactorial Research on Endogenous Psychoses

L. Demisch, F. Reinhuber, and H. J. Bochnik

Introduction

Evidence for the operation of genetic factors in the origin of major psychiatric diseases led to the suggestion that neurochemical variants might be identifiable in the cellular systems of patients suffering from these illnesses. The discovery of one or more of such genetic markers, susceptible to objective evaluation, would be of great theoretical and practical importance. As Kety recently pointed out, "it was such discoveries in the field of mental retardation which made possible the progress that has occurred in the delineation of homogenous subtypes, the elucidation of genetic transmission and their metabolic concomitants" (Kety 1981).

Some criteria for genetic markers in psychiatric disorders are shown in Table 1. Monoamine oxidase (MAO) in human blood platelets fulfills a number of these criteria. Platelets are readily accessible and repeated measurements of enzyme activity in patients and controls are possible. Measurements of platelet MAO activity allow conclusions about genetically stable characteristics of the individuals under study. Platelet MAO activity can be related to some neurophysiological variables and correlates with behavioral parameters. Increased or decreased activity of MAO in platelets, however, has not been specifically associated with any of the major psychoses (Murphy et al. 1974; DeLisi et al. 1982). Moreover, the original finding of a higher frequency of low MAO in platelets of chronic schizophrenics or manic-depressive patients has become a focus of disagreement. Furthermore, decreased MAO was found in platelets from subjects with other severe forms of psychopathology, including alcoholism and increased risk of suicide, indicating that altered platelet MAO may be a correlate of increased vulnerability to a broad range of chronic psychiatric disorders. As a consequence, studies were designed to search for a specific modus of psychopathogenesis which relates MAO to disordered behavior. These studies may be of relevance in the context of research addressed to the elucidation of heterogeneity of etiological factors in the major psychoses (Buchsbaum et al. 1980; Buchsbaum and Rieder 1979).

Table 1. Some criteria for genetic markers in psychiatry

Acessible, reliable, repeatable measurability
(Mode of) Genetic transmission
Relation to neurophysiological variables
Relation to behavioral parameters
Definite association with (a) psychiatric illness(es)

Biochemical Properties of MAO in Platelets

The localization of mitochondrial MAO activity in blood platelets of humans was described in 1964 (Paasonen et al. 1964). Blood platelets were subsequently used by different groups as an enzyme source in monitoring the efficacy of MAO inhibitors in humans (Palm et al. 1971). Collins and Sandler (1971) demonstrated that this enzyme was different from MAO in liver and placenta, and it is now well established that MAO in human blood platelets is of type B activity (Murphy and Donnelly 1974). Despite different interindividual MAO activities, comparable turnover numbers were measured in platelets from both patients and controls, indicating that MAO activity in platelets is directly proportional to the number of MAO molecules (Giller et al. 1982; Fowler et al. 1980; Summers et al. 1982). Platelets of different sizes have been shown to contain an equal number of mitochondria (Corash 1980). In addition, no correlation between the inner mitochondrial marker enzyme, succinate dehydrogenase, and platelet MAO activity could be found (Murphy et al. 1976). Therefore, it could be assumed that the approximately tenfold interindividual difference in MAO activity was due to a different number of MAO molecules on the outer mitochondrial membrane.

The process which regulates the number of MAO molecules on the outer mitochondrial membrane appears to be under a substantial degree of genetic control. Twin and pedigree studies have shown a high heritability of platelet MAO activity. About 25% (Gershon et al. 1980) to 70% (Reveley et al. 1983) of the variance could be explained by the contribution of genetic factors. The modus of genetic transmission is at present unknown. No study could confirm the modus of transmission controlled by either a single dominant or recessive gene locus. The activity of MAO in platelets is apparently affected by many genes and probably reflects the differential expression of genes controlling the processing of the enzyme, e.g., a structural gene for the active subunit, MAO degradation, availability of the flavin cofactor (Giller et al. 1982; Pintar and Breakefield 1982; Summers et al. 1982). MAO activity in platelets is unimodally distributed both in normal controls and in different patient populations (Wyatt et al. 1975; Gershon et al. 1980). In addition, no electrophoretic variance using nonequilibrium pH gradient electrophoresis was found and no differences between patients and controls were observed after the separation of ^{3}H-pargyline-labeled molecules, indicating no differences in the net charge of the molecules (Giller et al. 1982).

Platelets are cytoplasmic fragments of their parent megakaryocytes, which vary in ploidy values from 4 N to 64 N (N equals a haploid chromosome complement). and a correlation exists between platelet size and the replication of the megakaryocyte chromosomal complement (Caen et al. 1977). Therefore, the quality and quantity of transfer of genetic material to the platelet population may influence MAO activity, as has been suggested by Burch (1979). The 8-N megakaryocyte forms compact platelets, while 32 N produces large sponge-like platelets with a large surface-connecting system.

Human platelets in the peripheral circulation are heterogeneous for a number of characteristics, including cell volume, granular content, serotonin concentration and uptake, carbohydrate metabolism, and lipid peroxidation (Karpatkin 1969; Corash 1980). Platelet heterogeneity may be caused by variations in territory growth and membrane demarcation and/or by aging processes of platelets in the peripheral circulation (Penington 1981; Corash et al. 1979).

High MAO activity has been associated with heavy platelets in some but not all individuals (Friedhoff et al. 1978; Murphy et al. 1978). Any measurement of MAO, therefore, reflects the average enzyme value of the heterogeneous platelet population sampled at a particular time. Changes in the platelet population (due to severe exercise or noradrenaline infusions (Gawel et al. 1981) and reactive thrombocytosis in alcoholics (Demisch and Demisch 1982) have been associated with increases in the mean platelet MAO activity of the isolated platelet population. In studies comparing platelet MAO activity in psychiatric patients and in controls, it is therefore necessary to isolate representative platelet samples and to control long-term and short-term factors that might affect MAO activity (for a review see Demisch et al. 1983 a).

Relationship of Platelet MAO Activity to Neurophysiological Parameters

Most studies on platelet MAO make reference to MAO in the central nervous system, where MAO contributes to a major pathway in the degradation of monoamine neurotransmitters within neuronal and nonneuronal cells. In line with this reference, it is frequently presupposed that abnormal activity of MAO in platelets may reflect alterations in the control of amine levels at critical receptor sites in neurons, or may account for an accumulation of amines (e.g., dopamine, β-phenylethylamine) or abnormal amine metabolites (N,N-dimethylated indoleamines, etc.). The only study comparing MAO activity in four different parts of the brain after autopsy and in platelets from the same geriatric patients before death failed to demonstrate any correlation between the two enzyme activities (Winblad et al. 1979). It is important to consider, however, that blood platelets are „homogeneous" cell populations and a comparison of MAO activity between these cell fragments and MAO type A or B in brain tissue samples is difficult to make, since brain tissue contains a number of nonneural elements such as glia cells and blood vessels with high MAO activity.

Recently, Levitt and colleagues (1982) reported that MAO type B is specifically located in the adult rat brain in two major central nervous system cell classes, astrocytes and serotonin-containing neurons, whereas oligodendrocytes lack MAO-B activity. The authors discuss the important functional implications of MAO type B in glia cells that invest brain capillaries and circumventricular organs. Presumably, MAO type B activity in these structures may control, in part, the entrance of circulating biogenic amines into the brain. With respect to the considerable species differences in substrate preference and localization of MAO type A and B activity, a specific localization of MAO type B in the human brain appears to be necessary before definite conclusions about a relationship between platelet MAO type B activity and brain MAO type B activity can be drawn.

Table 2 summarizes the results from studies which correlated platelet MAO with neurophysiological parameters. These studies provide some evidence for a relationship between MAO in peripheral blood cells and processes in the brain expressed by serotonergic and/or dopaminergic activity. Oreland and colleagues (1981) reported a significant correlation between MAO activity in platelets and the concentration of 5-hydroxyindoleacetic acid (5-HIAA) and homovanillic acid (HVA). However, no correlation was found for the noradrenaline metabolite 3-methoxy-4-hydroxyphenylglycol (MHPG) in cerebrospinal fluid (CSF) from healthy volunteers. On the basis of postmortem studies on MAO activity and concentrations of 5-HIAA and serotonin in parts of the brain, they concluded that platelet MAO activity might reflect an "aspect" of brain serotonergic activity and proposed that the number of serotonergic neurons in the brain is different among individuals and that this difference is reflected by MAO activity in platelets (Oreland and Fowler 1979). This suggestion is of particular interest in view of the specific localization of MAO type B activity in central serotonergic neurons, as discussed above.

Using neuroendocrinological methods, Kleinmann et al. (1979) and Malas and colleagues (1983) proposed that platelet MAO activity may reflect an aspect of central nervous dopaminergic functioning. These authors measured significant negative correlations between apomorphine-induced growth hormone responses or α-methyl-p-tyrosine-induced prolactin increases and platelet MAO activity. In addition, basal prolactin concentrations in plasma were correlated with MAO activity in platelets from schizophrenic patients and normal subjects with low platelet MAO activity (Kleinmann et al. 1979). This finding could not be replicated, however, by Baron and colleagues (1983).

In addition, platelet MAO activity has been studied in relation to the amplitudes of visually evoked cortical potentials (VEP) and eye blinking rates. Figure 1 shows a linear regression analysis according to Pearson between MAO in platelets from 30 healthy subjects and amplitudes in VEPs (P_{100} component) of these subjects. The finding of a weak but significant correlation is in accordance with a study done by Von Knorring and colleagues on psychiatric patients (1977). Both studies failed to demonstrate a correlation between platelet MAO activity and the slope of stimulus

Table 2. Relationship of platelet monoamine oxidase activity to neurophysiological parameters

Subjects	Neurophysiological measure	Reference
Healthy volunteers ($n=42$)	5-HIAA; HVA in CSF	Oreland et al. 1981a
Schizophrenic patients ($n=10$)	0.75 mg Apomorphine; growth hormone response	Malas et al. 1983
Schizophrenic patients ($n=12$)	α-Methyl-p-tyrosine; plasma PRL increase	Kleinmann et al. 1979
Psychiatric patients ($n=57$)	Flash VEP; amplitude (N_1); latencies (2, 3, 4)[a]	Von Knorring et al. 1977
Healthy volunteers ($n=30$)	Flash VEP; amplitude (P_{100})	Reinhuber et al. 1983
Schizophrenic patients ($n=30$)	Eye blinking rate	Freed et al. 1980
Schizophrenic patients ($n=20$)	Eye blinking rate	Karson et al. 1983

5-HIAA, 5-hydroxyindoleacetic acid; HVA, homovanillic acid; PRL, prolactin; N_1, amplitude at first negative trough

[a] Latency from stimulas to first negative trough ($=2$), from stimulus to second positive peak ($=3$), and from stimulus to second negative trough ($=4$)

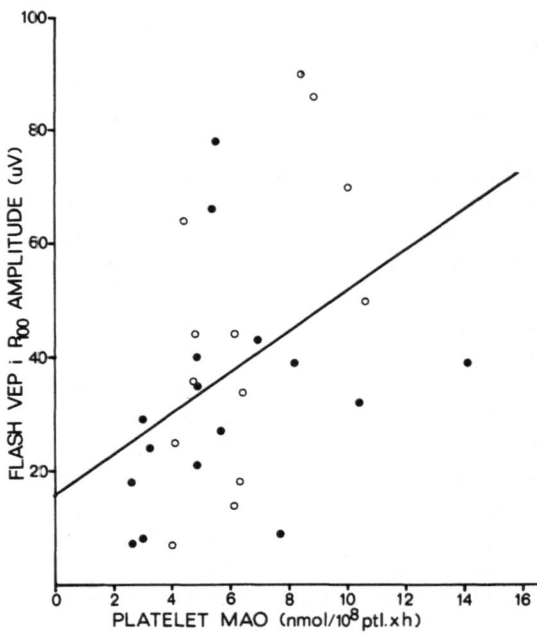

Fig. 1. Seventeen men and 13 women were selected from a group of 122 healthy subjects according to the lowest median and upper 10% of platelet MAO activity. Flashes were obtained by using a GRASS PS 22 photic stimulator (intensity: 6.9×10^6 cd/m^2; frequency: 2 flashes/s). Peak to trough amplitudes were measured. ●, men; ○, women. $r_{x,y} = 0.42$; $p < 0.05$ (Pearson product moment correlation, two tailed)

intensity/amplitude functions (e.g., the proneness to augment or reduce the intensity of the incoming signal with increasing stimulus intensity). Correlations between amplitudes in VEP and platelet MAO activity, however, reached significance only if flashes of low intensity were used, indicating that the setting of VEP measurement is of importance (Reinhuber et al. 1983). Therefore, subjects with decreased MAO in platelets may more likely show low amplitudes in the "middle" VEPs and increased eye blinking rates (Freed et al. 1980; Karson et al. 1983). Reduced amplitudes in the "middle" VEPs have been frequently measured in schizophrenic patients and are suggested to relate to attentional deficits in these patients (see Buchsbaum 1977).

Relationship of Platelet MAO Activity to Personality Characteristics

In their biochemical "high-risk" strategy, Buchsbaum and colleagues (1976) first reported about correlations between platelet MAO activity and personality characteristics. A number of independent studies have confirmed such a relationship. Low MAO activity in platelets from healthy volunteers or psychiatric patients has been linked to sensation seeking (Murphy et al. 1977; Schooler et al. 1978), monotony avoidance (Perris et al. 1980), thrill seeking and interest in mountaineering (Fowler et al. 1980) or extraversion (Gattaz and Beckmann 1981). More first-degree relatives with psychiatric hospitalizations or suicide attempts or problems with the law were found in healthy volunteers with low platelet MAO than in high platelet MAO probands (Buchsbaum et al. 1976). Healthy subjects with low platelet MAO and from different communities (urban, suburban, or rural) drank alcohol more frequently and smoked more cigarettes than high platelet MAO subjects (Buchsbaum and Coursey 1983; Demisch et al. 1982). Cigarette smoking and alcohol consumption had no acute effects on platelet MAO (Oreland et al. 1981b; Fowler et al. 1981) and chronic influences did not appear to account for the significant relationship between platelet MAO activity and personality characteristics (Fowler et al. 1980; Demisch et al. 1982).

High MAO activity in platelets was correlated with introversion in a group of psychiatric patients (Adler et al. 1980) and in a group of healthy volunteers (Demisch et al. 1982). In this latter study, high platelet MAO was associated with males who tended to be more irritable and hesitant, more passive and withdrawn, uncommunicative and less sociable (Demisch et al. 1982). A positive correlation between high MAO activity and high scores on the paranoia scale of the Minnesota Multipersonal Inventory (MMPI) was reported by Haier et al. (1979) in a group of males with extreme MMPI scores. These subjects, who were more concerned with themselves than with others, unable to cope with criticism, suspicious, or paranoid, were characterized as having symptoms related to affective disorders rather than to schizophrenia (Haier et al. 1979).

Calculations of multiple correlations using platelet MAO and VEP measures as predictor variables for test scores of personality factors indicated that subjects with high platelet MAO and high VEP amplitude stimulus intensity slopes ("augmenters") tended to be more irritable and hesitant (explained variance about 30%) (Reinhuber et al. 1983). In this sample, augmenters were found to be more emotionally labile. On the other hand, low MAO augmenters scored high on the MMPI schizophrenia scale (Coursey et al. 1979). All subjects in a group of psychiatric patients who were augmenters and had low MAO activities were found to be alcoholics (Von Knorring and Oreland 1978). More relatives from the low MAO augmenter group attempted suicide than from the low or high MAO activity reducer group (Buchsbaum et al. 1977).

If one attempts to summarize these studies on the relationship between platelet MAO and personality characteristics, it is important to recall that all studies reporting significant correlations between the two variables explain only minor proportions of variance. Furthermore, sampling a heterogeneous subject population has been demonstrated to be of importance for obtaining significant correlations (Haier et al. 1979). For example, Propping and colleagues (Propping and Friedl 1983) were unable to find any consistent, significant correlations between personality scores and platelet MAO in the study on medical students in an attempt to replicate Buchsbaum's original report on the biochemical high-risk paradigm. In addition, Reinhuber et al. (1983) were unable to replicate the association of platelet MAO activity with an extraversion/introversion dimension in a follow-up study on an urban population. The results of this comparison between correlations of platelet MAO and personality factors of a male sample from an urban and a rural community are shown in Table 3. In this sample, a significant negative correlation between platelet MAO and scores of the factor nervousness was computed, indicating that low MAO males are psychosomatically more disturbed than high MAO males. When the healthy subjects not on medication were combined from both samples,

Table 3. Platelet monoamine oxidase and personality factors in male controls

Factor FPI-A	Rural sample	Urban sample	Healthy males not on medication (both samples)
	$(n=54)$	$(n=71)$	$(n=70)$
Nervousness	–	– 0.27*	–
Depressivity	–	– 0.20	–
Composure	– 0.29*	–	– 0.25*
Extraversion	– 0.27*	–	–

* $p < 0.05$ Spearman rank two-tailed

high platelet MAO activity males were more irritable and hesitating. Apparently, there is no comprehensive hypothesis which predicts whether or not high or low platelet MAO activity is associated with premorbid personality traits and vulnerability for any of the major psychoses and which is sensitive to experimental evaluation. Keeping these limitations in mind, however, MAO in platelets appears to be more closely associated with personality traits which describe an affective psychopathology. This has also been emphasized by Buchsbaum and colleagues (Buchsbaum and Coursey 1983). Furthermore, this view also seems to be compatible with an association of alcoholism and suicidal risk with low platelet MAO activity.

Association of Platelet MAO Activity with Endogenous Psychosis

The results of decreased MAO in platelets of patients suffering from schizophrenic psychosis are inconsistent as are research findings on schizophrenic or affective psychoses. Basically, methodological problems appear to account for most of the controversy, along with the proposed biological heterogeneity of these disorders (Houlihan 1977). Long-term influences of neuroleptic treatment and problems in determining MAO activity turned out to be the most important methodological factors (for a summary see Friedhoff and Miller 1980; DeLisi et al. 1982; Demisch et al. 1983a). In general, there is agreement that some neuroleptics decrease MAO activity in platelets after an intake of longer than 14 days (Owen et al. 1981). If only those studies are considered in which neuroleptic treatment was omitted for a controlled period of time, it is evident that MAO activity in the platelets of those schizophrenic patients is not different from that in healthy controls (without neuroleptic medication for more than 3 months: Takahashi et al. 1975; Owen et al. 1976, 1981; Mann et al. 1979, 1981; DeLisi et al. 1982; without medication for more than 14 days: Friedhoff et al. 1978; Chojnacki et al. 1981; without neuroleptics for at least 7 days: Friedman et al. 1974; Jackman and Meltzer 1983).

It should be noted, however, that various types and doses of neuroleptics appear to affect platelet MAO activity in patients differently (e.g., chlorpromazine had no significant influence (Takahashi et al. 1975; Owen et al. 1981); perphenazine (10 mg/day) did not decrease platelet MAO after 3 weeks despite wide fluctuations in some individuals (A. Steiger, unpublished observation); fluphenazine produced an approximately 20% decrease after 3 weeks of treatment (Owen et al. 1981), haloperidol an approximately 20% decrease after 2 weeks of treatment, and butaperazine an approximately 40% decrease after 2 weeks of treatment (Chojnacki et al. 1981). Furthermore, Gattaz (1983) found decreased platelet MAO activity only in "phenothiazine-positive" males but not in females. The different effect of various neuroleptics in individuals treated over different time periods may explain the fact that some studies with neuroleptically treated patients did not find differences in platelet MAO activity compared with controls (Groshong et al. 1978; Oreland and Fowler 1979). Though most lines of evidence indicate that there is not a higher frequency of reduced MAO activity in platelets of chronic schizophrenic patients it should be kept in mind that Domino and Khanna (1976) reported significantly reduced platelet MAO activity in neuroleptic-free schizophrenic patients. Furthermore, Reveley et al. (1983) found significantly reduced (about 20%) platelet MAO in schizophrenic monozygotic twins compared to controls and also lower values in the psychiatrically well, neuroleptic-free twin, indicating that reduced platelet MAO may not be a result of the illness or its treatment.

On the basis of the suggested heterogeneity among schizophrenics and the controversial results on platelet MAO, the hypothesis was advanced that reduced enzyme activity may be typical only for subgroups of schizophrenic patients (Demisch et al. 1977). Initial studies suggested that the paranoid subgroup of schizophrenic patients is characterized by lower platelet MAO activity in comparison to controls or other undifferentiated schizophrenic patients (Demisch et al. 1977; Potkin et al. 1978). In addition, Potkin and colleagues (1979) reported that paranoid schizophrenics excreted higher amounts of β-phenylethylamine than nonparanoid patients. The finding that schizophrenic patients with a paranoid syndrome have significantly lower platelet MAO than undifferentiated chronic schizophrenics was not confirmed, however, by other studies (Groshong et al. 1978; Bond et al. 1979; Mann and Thomas 1979; Owen et al. 1981; for a review see DeLisi et al. 1982).

In another attempt it was questioned whether or not individuals within the schizophrenic or paranoid-schizophrenic patient group with deviant platelet MAO characteristics may differ with regard to important clinical features. In a pilot study by Demisch et al. (1983 b), MAO activities from 21 schizophrenic patients of the paranoid syndrome subtype did not show a unimodal distribution. A statistical test on a low MAO activity subgroup, however, failed to reach significance. The low and high platelet MAO patients within the paranoid-type schizophrenia group were separated at best by age of onset of first symptoms (below and above 25 years of age). As a result, we suggested that platelet MAO activity is lower only in a subgroup of patients with schizophrenia of the paranoid syndrome subtype, characterized by an early onset of symptoms (Demisch et al. 1983 b). A follow-up study on 64 paranoid-type schizophrenic patients (32 men, 32 women) did not, however, confirm this hypothesis (Demisch et al. 1984). In this study, the group of patients showed a unimodal distribution of platelet MAO activity values which was not different from a control group of 48 men and 62 women.

In addition to the subtype of schizophrenic patients with a paranoid syndrome, MAO activity was mesaured in a group of 31 schizoaffective patients (16 men and 15 women) referred to by us as having an atypic-phasic psychosis, which is characterized by a phasic course of the illness with complete remission of symptoms, resembling affective psychoses. Analysis of variance between controls, paranoid schizophrenic patients, and schizoaffective patients did not show any statistically significant differences between the groups if platelet MAO activity was expressed as nanomoles of deaminated products per 10^8 platelets per hour. It should be noted, however, that a significantly reduced (about 20%, $p < 0.05$) platelet MAO activity was measured in the patient groups compared to the controls if platelet MAO activity was expressed on a milligrams of protein basis (Demisch et al. 1984). This may explain the fact that in our first two studies (Demisch et al. 1977; Demisch 1982), in which platelet MAO activity was calculated on a milligrams of protein basis, we did find differences between schizophrenics and controls. If 25% of the bottom and top distribution of platelet MAO activity of males and females with a paranoid syndrome subtype of schizophrenia were compared with respect to age of onset of first symptoms, duration of illness, and number of hospitalizations, no differences could be computed. Comparable results were obtained after dichotomizing the patient groups by the median. A similar negative result was found when the same approach was used for the schizoaffective patient group. A higher frequency of psychiatric disorders, suicide, or alcoholism was not found in the first-degree relatives of the low platelet MAO patient group, either for schizophrenics or for schizoaffective patients. If all of the paranoid schizophrenic patients with an early onset of symptoms (below 25 years of age) and a definite case of a psychiatric disorder in first-degree relatives were compared with patients showing a record of no familial occurrence of psychiatric disorders and a later onset of symptoms, there was no difference in platelet MAO activity between the two groups. The results of this study on a "homogeneous" group of schizophrenic patients, characterized by a regular occurrence of paranoid symptoms, do not support the hypothesis that low platelet MAO activity is a significant factor in differentiating subgroups of patients, either with respect to different symptomatologies or with regard to clinically relevant descriptive data. Therefore, the suggestion that platelet MAO activity is useful in studies on the heterogeneity of etiological factors in endogenous psychosis cannot be empirically confirmed to date. This finding is in accordance with similar results from other studies on undifferentiated schizophrenic patients (see Wyatt et al. 1979; Owen et al. 1981).

At present, platelet MAO has not been established as a predictor of increased risk for a schizophrenic or affective psychosis. Gershon et al. (1980) and Maubach et al. (1981) concluded that reduced MAO does not function as a genetic marker of vulnerability to affective psychosis, although the former study, in contrast to the latter one, did find differences between the bipolar depressive patient group and controls. The hypothesis that decreased MAO may represent a genetic predisposition for the development of schizophrenia, as originally derived from studies on monozygotic twins discordant for schizophrenia (Wyatt et al. 1973), has not been supported by the evidence to date (for a summary see DeLisi et al. 1982). The only published study on a multigenerational pedigree analysis of schizophrenic families reported that the diagnosed schizophrenics had lower MAO activities than those of the other family members (Book et al. 1978). Additional careful pedigree analysis or studies on families with high densities of chronic schizophrenia seem to be necessary before definite conclusions associating platelet MAO activity with a genetic predisposition can be drawn.

The overall impression derived from the clinical data indicates that platelet MAO activity does not provide a reliable basis for confirming hypotheses about the heterogeneity of schizophrenic psychoses or describing particular subgroups of patients with an increased risk of endogenous psychosis.

Summary

Blood platelet MAO activity has a number of useful features for multifactorial research on endogenous (major) psychoses. An attempt is made to summarize the biochemical properties of this enzyme, the relationship of platelet MAO to brain physiology, the correlates of this enzyme with behavioral characteristics, and its association with endogenous psychoses. The overall impression indicates that MAO activity in the peripheral blood cells reflects an aspect of serotonergic or dopaminergic functions in the brain and it appears to be associated with personality traits connected especially with affective symptoms (e.g., irritability, nervousness). Correlations between platelet MAO and different neurophysiological or psychological variables explain, however, only minor proportions of variance. As a result, no comprehensive and experimentally testable hypothesis is at present able to link high or low platelet MAO activity to schizophrenic or affective psychoses. The clinical data did not show a definite association of reduced platelet MAO activity with chronic schizophrenia. In addition, the literature on increased or decreased MAO in platelets from patients with different forms of depression is inconsistent. Furthermore, platelet MAO has not been established as a predictor of increased risk for any of the endogenous psychoses. Attempts to search for a specific modus of psychopathogenesis relating MAO to disordered behavior have not so far been successful in describing particular subgroups of patients (e.g., paranoid schizophrenics). Finally, the suggestion that platelet MAO activity may be useful for studies on the heterogeneity of etiological factors in endogenous psychosis cannot be empirically confirmed to date. Further progress in clinical studies using platelet MAO as a variable needs more basic knowledge about the modus of genetic inheritance of MAO in platelets and in the brain, including platelet-specific mechanisms in processing the enzyme and the function of MAO type B in different cells of the brain. Despite increasing knowledge of basic biochemical and neurophysiological processes, however, the intrinsic problem of elucidating multifactorial systems such as psychiatric diseases remains: multiple and significant partial correlations explaining minor proportions of variance do not allow an unequivocal classification and prognosis in the individual case.

Aknowledgments. We would like to thank Dr. G. Buchholz, Dr. K. Demisch, K. Georgi, G. Gebhart, P. Kaczamrczyk, and S. Guttzeit for their cooperation. This study was supported by a grant of the DFG (Bo 211/212).

References

Adler SA, Gottesman II, Orsula PJ, Kizuka PP, Schildkraut JJ (1980) Platelet MAO activity: relationships to clinical and psychometric variables. Schizophr Bull 6:226–231

Baron M, Levitt M, Asnis L, Fein M (1983) Prolactin levels in schizophrenia: relation to platelet monoamine oxidase, plasma amine oxidase, plasma dopamine-beta-hydroxylase and erythrocyte catechol-O-methyltransferase activity. Biol Psychiatry 18:579–582

Bond PA, Cundall RL, Falloon IRH (1979) Monoamine oxidase (MAO) of platelets, plasma, lymphocytes and granulocytes in schizophrenia. Br J Psychiatry 134:360–365

Book JA, Wetterberg L, Modryewska K (1978) Schizophrenia in a North Swedish geographical isolate 1900–1977: epidemiology, genetics and biochemistry, Clin Gen 14:373–394

Buchsbaum MS (1977) The middle evoked response components and schizophrenia. Schizophrenia Bull 3:93–104

Buchsbaum MS, Coursey RD (1983) Biological high-risk paradigm and platelet MAO activity in community samples. In: Beckmann H, Riederer P (eds) Monoamine oxidase and its selective inhibitors. Karger, Basel pp 278–286 (Modern problems of pharmacopsychiatry, vol 19)

Buchsbaum MS, Rieder RO (1979) Biological heterogeneity and psychiatric research. Arch Gen Psychiatry 36:1163–1169

Buchsbaum MS, Coursey RD, Murphy DL (1976) The biochemical high-risk paradigm: behavioral and familial correlates of low platelet monoamine oxidase activity. Science 194:339–341

Buchsbaum MS, Haier RJ, Murphy DL (1977) Suicide attempts, platelet monoamine oxidase and the average evoked response. Acta Psychiatr Scand 56:69

Buchsbaum MS, Coursey RD, Murphy DL (1980) Schizophrenia and platelet monoamine oxidase: research strategies. Schizophr Bull 6:375–384

Burch EA (1979) Platelet MAO research: complications. Am J Psychiatry 136:237

Caen JP, Cronberg S, Kubisz P (eds) (1977) Platelets: physiology and pathology. Stratton, New York

Chojnacki M, Kralik P, Allen RH, Ho BT, Schoolar JC, Smith RC (1981) Neuroleptic-induced decrease in platelet MAO activity of schizophrenic patients. Am J Psychiatry 138:838–840

Collins GGS, Sandler M (1971) Human blood platelet monoamine oxidase. Biochem Pharmacol 20:289–296

Corash L (1980) Platelet heterogeneity: relevance to the use of platelets study psychiatric disorders. Schizophr Bull 6:254–258

Corash L, Shafer B, Perlow M (1979) Heterogeneity of human whole blood platelet subpopulations. II. Use of sub-human primate model to analyze the relationship between density and platelet age. Blood 52:726–734

Coursey RD, Buchsbaum MS, Murphy DL (1979) Platelet MAO activity and evoked potentials in the identification of subjects biologically at risk for psychiatric disorders. Br J Psychiatry 134:372–381

DeLisi LE, Wise CD, Bridge TP, Phelps BH, Potkin SG, Wyatt RJ (1982) Monoamine oxidase and schizophrenia. In: Usdin E, Hanin I (eds) Biological markers in psychiatry and neurology. Pergamon, Oxford, pp 79–134

Demisch K, Demisch L (1982) Stimulation von Monoamin Oxidase and Adenylat-Cyclase in Blutplättchen von Patienten im Verlauf von Alkoholentzugsdeliren. In: Beckmann H (ed) Biologische Psychiatrie. Thieme, Stuttgart, pp 168–175

Demisch L (1982) Die Bedeutung veränderter Monoamin Oxidase bei Patienten mit psychiatrischen Erkrankungen. Nervenarzt 8:455–460

Demisch L, Von der Mühlen H, Bochnik HJ, Seiler N (1977) Substrate-typic changes of platelet monoamine oxidase activity in subtypes of schizophrenia. Arch Psychiatr Nervenkr 224:319–329

Demisch L, Georgi K, Patzke B, Demisch K, Bochnik HJ (1982) Correlation of blood platelet MAO activity with introversion: a study on a German rural population. Psychiatry Res 6:303–311

Demisch L, Kaczmarczyk P, Gebhart P (1983a) Methodological issues using platelet MAO in psychiatric research. In: Beckmann H, Riederer P (eds) Monoamine oxidase and its selective inhibitors. Karger, Basel, pp 265–277 (Modern problems of pharmacopsychiatry, vol 19)

Demisch L, Demisch K, Von der Mühlen H (1983b) Endogene Psychosen und veränderter Monoaminoxidase-Aktivität: Blutplättchen als periphere Modellsysteme für biochemischpsychiatrische Untersuchungen. In: Wanke K, Richtberg W (eds) Erlebte Psychiatrie. Perimed, Erlangen, pp 183–191

Demisch L, Buchholz G, Demisch K, Bochnik HJ (1984) Platelet MAO in paranoid schizophrenic and schizoaffective patients.

Domino EF, Khanna SS (1976) Decreased platelet MAO activity in unmedicated chronic schizophrenic patients. Am J Psychiatry 133:323–326

Fowler CJ, Von Knorring L, Oreland L (1980) Platelet monoamine oxidase activity in sensation seekers. Psychiatry Res 3:273–279

Fowler CJ, Wiberg A, Oreland L, Danielsson A, Palm U, Winblad B (1981) Monoamine oxidase activity and kinetic properties in platelet-rich plasma from controls, chronic alcoholics and patients with nonalcoholic liver disease. Biochem Med 25:356–365

Freed WJ, Kleinmann JE, Karson CN, Potkin SG, Murphy DL, Wyatt RJ (1980) Eye-blink rates and platelet monoamine oxidase activity in chronic schizophrenic patients. Biol Psychiatry 15:329–332

Friedhoff AJ, Miller JC (1980) Platelet monoamine oxidase as a function of nongenetic factors. Schizophr Bull 6:314–319

Friedhoff AJ, Miller JC, Karpatkin S (1978) Heterogeneity of human platelets. VII: Platelet monoamine oxidase in normals and patients with autoimmune thrombocytopenic purpurea and reactive thrombocytosis: its relationship to platelet protein density. Blood 51:317–323

Friedman E, Shopsin B, Sathananthan G, Gershon S (1974) Blood platelet monoamine oxidase activity in psychiatric patients. Am J Psychiatry 131:1392–94

Gattaz WF (1983) Platelet MAO activity in major psychoses. In: Beckmann H, Riederer P (eds) Monoamine oxidase and its selective inhibitors. Karger, Basel, pp 315–320 (Modern problems of pharmacopsychiatry, vol 19

Gattaz WF, Beckmann H (1981) Platelet MAO activity and personality characteristics in schizophrenic patients and normal individuals. Acta Psychiatr Scand 63:479–485

Gawel MJ, Glover V, Burkitt M, Sandler M, Rose FC (1981) The specific activity of platelet monoamine oxidase varies with platelet count during severe exercise and noradrenaline infusion. Psychopharmacology 72:275–277

Gershon ES, Goldin LR, Lake CR, Murphy DL, Guroff JJ (1980) Genetics of plasma dopamine-β-hydroxylase (DBH), erythrocyte catechol-O-methyltransferase (COMT) and platelet monoamine oxidase (MAO) in pedigrees of patients with affective disorders. In: Usdin E, Souches P, Youdim MBH (eds) Enzymes and neurotransmitters in mental disease. Wiley, London, pp 281–299

Giller E, Castiglione C, Wojciechowski J, Breakefield XO (1982) Molecular properties of platelet MAO in psychiatric patients and controls. In: Usdin E, Hanin I (eds) Biological markers in psychiatry and neurology. Pergamon, Oxford, pp 111–134

Groshong R, Baldessarini RJ, Gibson A, Lipinski JF, Axelrod D, Pope A (1978) Activities of types A and B MAO and catechol-O-methyltransferase in blood cells and skin fibroblasts of normal and chronic schizophrenic subjects. Arch Gen Psychiatry 35:1198–1205

Haier RJ, Murphy DL, Buchsbaum MS (1979) Paranoia and platelet MAO in normals and nonschizophrenic psychiatric groups. Am J Psychiatry 136:308–310

Houlihan JP (1977) Heterogeneity among schizophrenic patients: selective review of recent findings (1970–75). Schizophr Bull 3:246–258

Jackmann HL, Meltzer HY (1983) Kinetic constants of platelet monoamine oxidase in schizophrenia. Am J Psychiatry 140:1044–1047

Karpatkin S (1969) Heterogeneity of human platelet I. Metabolic and kinetic evidence suggestive of young and old platelets. J Clin Invest 48:1073

Karson CN, Kleinman JE, Berman KF, Phelps BH, Wise CD (1983) An inverse correlation between spontaneous eye-blink rate and platelet monoamine oxidase activity. Br J Psychiatry 142:43

Kety SS (1981) Problems of genetic research on psychiatric illness. In: Gershon ES, Matthysse S, Breakefield XO, Ciaranello RD (eds) Genetic research strategies in psychobiology and psychiatry. Elsevier, Amsterdam, pp 395–397 (Psychobiology and psychopathology, vol 1)

Kleinman JE, Potkin S, Rogol A, Buchsbaum MS, Murphy DL, Gillin JC, Nasrallah HA, Wyatt RJ (1979) A correlation between platelet monoamine oxidase activity and plasma prolactin concentrations in man. Science 206:479–481

Levitt P, Pintar JE, Breakefield XO (1982) Immunocytochemical demonstration of monoamine oxidase B in brain astrocytes and serotonergic neurons. Proc Natl Acad Sci USA 79:6385–6389

Malas KL, Van Kammen DP, De Fraites EA, Brown GM, Gold PW (1983) Platelet monoamine oxidase and the growth hormone response of apomorphine in schizophrenia. Biol Psychiatry 18:255–259

Mann J, Thomas KM (1979) Platelet monoamine oxidase activity in schizophrenia. Relationship to disease, treatment, institutionalization and outcome. Br J Psychiatry 134:366–371

Mann JJ, Kaplan RD, Georgotas A, Friedman E, Branchey M, Gershon S (1981) Monoamine oxidase activity and enzyme kinetics in three subpopulations of density-fractionated platelets in chronic paranoid schizophrenics. Psychopharmacology 74:344–348

Maubach M, Diebold K, Friedl W, Propping P (1981) Platelet MAO activity in patients with affective psychosis and their first-degree relatives. Pharmacopsychiatry 14:87–93

Murphy DL, Donnelly CH (1974) Monoamine oxidase in man: enzyme characteristics in platelets, plasma and other human tissues. In: Usdin E (ed) Advances in biochemical psychopharmacology. Raven, New York, pp 71–85

Murphy DL, Belmaker R, Wyatt RJ (1974) Monoamine oxidase in schizophrenia and other behavior disorders. J Psychiatr Res 11:221–247

Murphy DL, Donnelly CH, Miller L, Wyatt RJ (1976) Platelet monoamine oxidase in chronic schizophrenia: some enzyme characteristics relevant to reduced activity. Arch Gen Psychiatry 33:1377–1381

Murphy DL, Belmaker R, Carpenter WT, Wyatt RJ (1977) Monoamine oxidase in chronic schizophrenia: studies of hormonal and other factors affecting enzyme activity. Br J Psychiatry 130:151–8

Murphy DL, Costa JL, Shafer B, Corash B (1978) Monoamine oxidase activity in different density gradient fractions of human platelets. Psychopharmacology 59:193–197

Oreland L, Fowler CF (1979) The activity of human brain and thrombocyte monoamine oxidase (MAO) in relation to various psychiatric disorders. I MAO activity in some disease states. In: Singer TP, Von Korf RW, Murphy DL (eds) Monoamine oxidase, structure, function and altered functions. Academic, New York, p 389

Oreland L, Wiberg A, Asberg M, Traskman L, Sjostrand L, Thoren P, Bertilsson L, Tybring G (1981a) Platelet MAO activity and monoamine metabolites in cerebrospinal fluid in depressed and suicidal patients and in healthy controls. Psychiatry Res 4:21–29

Oreland L, Fowler CJ, Schalling D (1981b) Low platelet monoamine oxidase activity in cigarette smokers. Life Sci 29:2511–2518

Owen F, Bourne R, Crow TJ (1976) Platelet monoamine oxidase in schizophrenia. Arch Gen Psychiatry 33:1370–1373

Owen F, Bourne RC, Crow TJ, Fadhli AA, Johnstone EC (1981) Platelet monoamine oxidase activity in acute schizophrenia – relationship to symptomatology and neuroleptic medication. Br J Psychiatry 139:16–22

Paasonen MK, Solatunturi E, Kivalo E (1964) Monoamine oxidase activity of blood platelets and their ability to store 5-hydroxytryptamine in some mental deficiencies. Psychopharmacology 6:120–124

Palm D, Fengler HJ, Güllner HG, Planz G, Quiring B, May D, Helmstaedt B, Lemmer B, Moon HK, Holler C (1971) Quantitation of irreversible inhibition of monoamine oxidase in man. Eur J Clin Pharmacol 3:82–92

Penington DG (1981) Formation of platelets. In: Gordon JL (ed) Platelets in biology and pathology 2. Elsevier, Amsterdam, pp 19–42

Perris C, Jacobsson L, Von Knorring L, Oreland L, Perris H, Ross SB (1980) Enzymes related to biogenic amine metabolism and personality characteristics in depressed patients. Acta Psychiatr Scand 61:477–484

Pintar JE, Breakefield XO (1982) Monoamine oxidase (MAO) activity as a determinant in human neurophysiology. Behav Genet 12:53–68

Potkin SG, Cannon HE, Murphy DL, Wyatt RJ (1978) Are paranoid schizophrenics biologically different from other schizophrenics? N Engl J Med 298:61–66

Potkin SG, Karoum F, Chuang LW, Cannon-Spoor HE, Phillips I, Wyatt RJ (1979) Phenylethylamine in paranoid chronic schizophrenia. Science 206:470–471

Propping P, Friedl W (1983) Platelet MAO activity and high risk for psychopathology in a German population. In: Beckmann H, Riederer P (eds) Monoamine oxidase and its selective inhibitors. Karger, Basel, pp 304–314 (Modern problems of pharmacopsychiatry, vol 19)

Reinhuber E, Demisch L, Berdjis H, Georgi K, Diehl R (1983) Visually evoked potential (flash and pattern stimuli) correlates of platelet MAO activity and personality characteristics. Neuro Sci Lett [Suppl] 14:304

Reveley MA, Reveley AD, Clifford CA, Murray RM (1983) Genetics of platelet MAO activity in discordant schizophrenic and normal twins. Br J Psychiatry 142:560–565

Schooler C, Zahn TC, Murphy DL, Buchsbaum MS (1978) Psychological correlates of monoamine oxidase activity in normals. J Nerv Ment Dis 166:177–186

Summers KM, Brown GK, Craig IW, Littlewood J, Peatfield R, Glover V, Rose FC, Sandler M (1982) Platelet monoamine oxidase: specific activity and turnover number in headache. Clin Chim Acta 121:139–146

Takahashi S, Yamane H, Tani N (1975) Reduction of blood platelet monoamine oxidase activity in schizophrenic patients on phenothiazines. Folia Psychiatr Neurol Jpn 29:209–214

Von Knorring L, Oreland L, Perris C (1977) Neurophysiological measures and visual averaged evoked responses in psychiatric patients. Relationship to MAO activity in platelets. Neuropsychobiology 3:65–74

Von Knorring L, Oreland L (1978) Visual averaged evoked responses and platelet monoamine oxidase activity as an aid to identify a risk group for alcoholic abuse. A preliminary study. Prog Neuropsychopharmacol 2:385–392

Winblad B, Gottfries CG, Oreland L, Wiberg A (1979) Monoamine oxidase in platelets and brains of non-psychiatric and nonneurological geriatric patients Med Biol 57:129–136

Wyatt RJ, Murphy DL, Belmaker R, Cohen S, Donnelly CH, Pollin (1973) Reduced monoamine oxidase activity in platelets. A possible genetic marker for vulnerability to schizophrenia. Science 173:916–918

Wyatt RJ, Belmaker R, Murphy DL (1975) Low platelet monoamine oxidase and vulnerability to schizophrenia. In: Mendlewicz J (ed) Genetics and psychopharmacology. Karger, Basel, pp 38–56 (Modern problems of pharmacopsychiatry, vol 10)

Wyatt RJ, Potkin SG, Murphy DL (1979) Platelet monoamine oxidase activity in schizophrenia: review of the data. Am J Psychiatry 136:377–382

Neuroendocrine Responses to Serotonin Agonists as Possible Markers of the Functional State of Serotonergic Neurotransmission in Psychiatric Disorders

E. A. MUELLER, L. J. SIEVER, and D. L. MURPHY

Introduction

An alteration in central serotonin (5-HT) neurotransmitter activity has been implicated in the pathogenesis of affective illness for over two decades. One formulation of the 5-HT hypothesis of depression postulates a direct association between reduced central 5-HT activity and depression (Van Praag 1962; Coppen et al. 1963; Lapin and Oxenkrug 1969). Evidence supporting this hypothesis has been mixed (Murphy et al. 1978; Abrams 1978) and has relied primarily on measurements of serotonin and its metabolites in blood and CSF and the response of depressives to administration of drugs thought to act by enhancing or reducing serotonergic function. Although much evidence exists supporting a relationship between reductions in 5-HT activity and depression, other evidence from behavioral studies and biochemical data in experimental animals has led to alternative hypotheses on the possible role of 5-HT function in affective illness (Aprison et al. 1978; Ogren et al. 1979).

More recently, studies in depressives measuring 5-HT uptake (Stahl et al. 1983) and ^{3}H-imipramine binding in platelets (Paul et al. 1981; Briley et al. 1980), as well as postmortem studies of 5-HT receptor numbers and ^{3}H-imipramine binding in the brains of suicide victims (Stanley et al. 1982; Stanley and Mann 1983), have provided additional support for the existence of an alteration in 5-HT function in depression. Likewise, the emergence of preclinical data from biochemical, behavioral, and electrophysiological studies of the effects of chronic antidepressant drug administration on neurotransmitter receptor properties, suggesting that alterations in 5-HT receptors may be important to the mechanism of action of effective antidepressant treatments (Anderson 1983; Fuxe et al. 1983), has added further support to the hypothesis that serotonergic mechanisms may play an important role in the pathogenesis of certain affective symptoms or syndromes.

Although studies such as these provide suggestive evidence for the existence of an alteration in the 5-HT system in depression, the strategies on which these studies are based provide little information on the functional activity of brain 5-HT system in the depressed state. An alternative approach, the pharmacologic challenge strategy, a paradigm which has been largely neglected in the study of 5-HT systems in man, may provide a more suitable method for the study of dynamic aspects of neurotransmitter activity in man. This approach takes advantage of the fact that the secretion of hypothalamic and pituitary hormones is regulated in part by the activity of central neurotransmitter systems (Martin et al. 1977). By virtue of this fact, the measurement of changes in the concentration of hormones produced by the administration of drugs with selective actions on different neurotransmitter systems may prove useful in providing an indication of the overall functional status of the neurotransmitter system of interest. In addition, since pharmacologic probes can be chosen not only to target particular transmitter systems, but also to perturb a given system at different points of regulation, the pharmacologic challenge strategy may also be useful in efforts to look for changes in receptor sensitivity which have been suggested to be of importance in the pathogenesis of certain psychiatric disorders and in the mechanism of action of antidepressant drugs (Bunney et al. 1977; Olsen et al. 1980).

In this paper, we will briefly present recently reported results (Siever et al. 1983) from ongoing studies designed to assess the net responsivity of the 5-HT system in depressed patients on the basis of changes in serum prolactin following administration of the serotonergic neurotransmitter system agonist fenfluramine, as an example of the potential usefulness of the pharmacologic challenge approach to assessment of serotonergic function in man. Since other serotonin agonists, especially its precursors 5-hydroxytryptophan (5-HTP) and L-tryptophan, are also currently being used with the same intent (Meltzer et al. 1983 a, b; Charney et al. 1982; Westenberg et al. 1982), and since our group has recently begun clinical studies using the serotonin receptor agonist m-chlorphenylpiperazine and the serotonin receptor antagonist metergoline to examine more discrete functional elements of the 5-HT synaptic complex in psychiatric patients (E. A. Mueller, D. L. Murphy, T. Aloi, and T. Insel, 1984, in preparation), attention will also be focused on issues influencing the validity of these various approaches to the assessment of central 5-HT function and on some of the problems encountered in using the pharmacologic challenge strategy for this purpose. In view of the growing evidence that biological changes associated with psychiatric dysfunction may serve as "markers" for psychiatric disorders (Bunney 1983; Usdin and Hanin 1982), such measures as serotonin-agonist-induced neuroendocrine changes possess potential diagnostic and therapeutic implications, and also promise to be useful in the study of hypotheses regarding the biochemical mechanisms underlying psychiatric disease states and the mode of action of thymoleptic drugs, provided certain limitations are kept in mind.

An Example of a Pharmacologic Challenge Strategy for the Serotonergic System: Neuroendocrine Responses to Fenfluramine

Prolactin Responses to Fenfluramine as an Index of Central Serotonergic Function

Fenfluramine, an anorectic drug used in the treatment of obesity, is a halogen-substituted phenylethylamine derivative which increases central serotonin activity. Biochemical data indicate that the 5-HT agonist properties of fenfluramine derive from its combined ability to promote a rapid release of 5-HT (Trulson and Jacobs 1976; Kannengiesser et al. 1976; Cleinschmidt and McGuffin 1978) and to inhibit 5-HT reuptake (Garattini et al. 1975). In addition, receptor binding studies suggest that both fenfluramine and its principal metabolite, norfenfluramine (NF), may have direct, albeit weak, 5-HT receptor agonist activity (Garattini et al. 1979).

Neuroendocrine studies have shown that acute fenfluramine administration is associated with an increase in serum prolactin in both experimental animals (Fuller et al. 1976) and man (Slater et al. 1976; Chase et al. 1976). This effect, which has been shown to occur by a mechanism involving 5-HT release (Quattrone et al. 1978), is consistent with the findings of a large body of work in both animals and man, suggesting that enhancement of 5-HT activity increases serum prolactin levels (Preziosi 1983; Wirz-Justice et al. 1976; Puhrringer et al. 1976; Lancranjan et al. 1977; MacIndoe and Turkington 1973; Charney et al. 1982; Kato et al. 1974; Woolf and Lee 1977). In addition, chemical lesioning experiments using the 5-HT neurotoxin 5,7-dihydroxytryptamine have been shown to produce serotonin receptor supersensitivity and to result in enhanced prolactin response to the serotonin agonist 5-methoxydimethyltryptamine (Kuhn et al. 1981), suggesting that 5-HT receptors mediating the prolactin rise might also manifest alterations in sensitivity as a result of fluctuations in endogenous levels of synaptic 5-HT postulated to exist in patients with depression. In view of these considerations, studies were initiated in depressed patients and controls measuring the prolactin response following oral administration of fenfluramine as an index of net responsiveness of the 5-HT system.

Subjects and Methods

The subjects in this study consisted of 18 medically healthy, nonobese psychiatric inpatients (13 females, five males; mean age 41.5 ± 2.9) and ten medically healthy nonobese controls (eight females and two males) selected to match ten of the patients with regard to sex, age to within 5 years, and menopausal status (mean age of ten patients 43.5 ± 3.8; mean age of controls 44.0 ± 4 years). All subjects gave written consent before being admitted to the study. All 18 patients met the Research Diagnostic Criteria (RDC) for major depressive disorder (six unipolar, 12 bipolar) and were depressed at the time of testing.

All subjects were drug free for a minimum of 2 weeks prior to testing. Plasma prolactin levels were measured by radioimmunoassay, with an overall intraassay reliability of 3.3% and an interassay reliability of 12.4%. Baseline plasma prolactin values were calculated as the average of the -15 and 0 min time points, and the prolactin response plateau, which began at 3 h following the oral administration of 60 mg fenfluramine for both patients and controls, was determined by averaging the prolactin values at 180, 240, and 300 min after drug administration.

Prolactin Responses to Fenfluramine in Depressed Patients and Controls

As summarized in Table 1, both the absolute increase ($p < 0.05$) and the percentage increase ($p < 0.01$) in plasma prolactin following fenfluramine administration was significantly smaller in the 18 depressed patients than the controls. Baseline prolactin was significantly higher in the depressed group than in controls ($p < 0.01$),

Table 1. Prolactin response to fenfluramine in depressed patients and normal controls

	Plasma prolactin (ng/ml)			
	Age (years)	Baseline	Increase after fenfluramine	Percentage increase after fenfluramine
Normal controls ($n = 10$)	44.0±4.0	8.5±1.1	17.2±3.3	248±58
All depressed patients ($n = 18$)	41.5±2.9	17.8±2.5*	10.9±2.7*	63±13**
Depressed patients excluding those with elevated (> 25 ng/ml) baselines ($n = 13$)	42.8±3.4	12.6±1.7	6.4±2.3*	58±16**
Depressed patients, age- and sex-matched with controls ($n = 10$)	43.5±3.8	15.5±2.4	6.8±2.9**	52±20**

* $p<0.05$, * $p<0.01$, Student's t-test compared to controls; means±SEM

principally as a result of the fact that a subgroup of five depressed patients showed an elevated baseline prolactin of greater than 25 mg/ml. When this subgroup of depressives with elevated baseline was excluded and the data reanalyzed, baseline prolactin values for the remaining group of depressives ($n = 13$) were no longer significantly different from controls, but the absolute increase in plasma prolactin among the remaining 13 depressives was still significantly reduced compared to controls ($p < 0.01$) (Table 1). Furthermore, when ten of the depressed patients were matched by sex and age (to within 5 years) with the ten controls and the comparisons reanaylzed, both the absolute and percent age fenfluramine-induced increase in prolactin remained significantly less in the depressed patients than in the control group (Table 1). This difference also remained significant ($p < 0.01$) when one of the ten patients with an elevated baseline and her paired control were excluded.

Baseline prolactin concentrations were not significantly correlated with prolactin elevations following fenfluramine in the entire group of 28 subjects ($r=0.25$, $p=$ NS), in the control-matched subgroup of ten depressed patients ($r=0.30$, $p=$ NS), or in the control subjects ($r=0.00$, $p=$ NS).

Figure 1 shows the increase in prolactin following fenfluramine for individual patients with their matched controls. Of particular interest is the fact that seven of the nine depressed patients had a minimal or nonexistent increase in prolactin (< 5 ng/ml), while in contrast only one of the nine matched controls had an increase of less than 8 ng/ml.

An Interpretation and Critique of the Fenfluramine-Induced Prolactin Differences Between Depressed Patients and Controls

The results of our initial study demonstrated that fenfluramine's ability to increase serum prolactin was significantly diminished in depressed patients compared to control subjects. One interpretation of this finding is that the diminished prolactin responsivity could reflect a diminished availability of serotonin for release by fen-

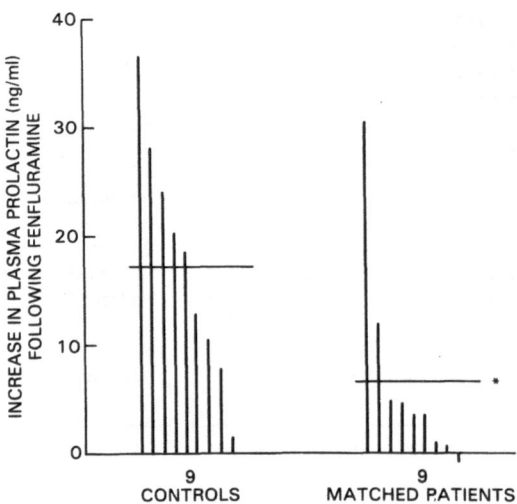

Fig. 1. Plasma prolactin increases from baseline in depressed patients and sex- and age-matched controls following 60 mg fenfluramine given orally. Means for patients (6.5 ± 3.2) and controls (18.2 ± 3.4) are indicated by the *horizontal lines*. *$p < 0.01$, Student's t-test

fluramine among depressives, an interpretation which would be consistent with the hypothesis of decreased serotonin availability in depression based on reports of decreased serotonin metabolite 5-hydroxyindoleacetic acid (5-HIAA) in the CSF, decreased accumulation of 5-HIAA in the CSF of probenecid-pretreated depressives, and reports of reduced serotonin in the brains of suicide victims (Van Praag 1981; Murphy et al. 1978).

One issue that could limit the validity of this interpretation is the baseline difference in prolactin concentrations between patients and controls. However, when patients with elevated baseline prolactin were excluded and the data reexamined, the prolactin response in depressives remained significantly diminished. Additionally, direct evaluation of correlations between baseline prolactin and fenfluramine-induced prolactin elevations show no suggestion of a negative relationship. Furthermore, other stimuli [i.e., insulin, thyrotropin-releasing hormone (TRH)] are capable of inducing marked increases in prolactin above the basal prolactin levels observed in this group of patients, and there is no evidence for a refractoriness to further stimulation at the baseline prolactin concentrations noted here (Perez-Lopez et al. 1981; Caufriez et al. 1981).

Pharmacologic Considerations Regarding Fenfluramine's Neuroendocrine Effects in Patients and Controls

Persuasive evidence exists suggesting that the increase in prolactin following fenfluramine administration occurs principally by a mechanism involving the release of 5-HT (Quattrone et al. 1978; Fuller et al. 1982). Although fenfluramine affects the dopaminergic system in rodents at high doses (i.e., above 5 mg/kg) (Pinder et al. 1975), such effects seem unlikely at the low doses (60 mg, i.e., less than 1 mg/kg) used in this study, since a higher dose (120 mg) of fenfluramine has been shown to alter the probenecid-induced accumulation of the serotonin metabolite, 5-HIAA without a significant change in the accumulation of the dopamine metabolite homovanillic acid (HVA) in another patient group (Shoulson and Chase 1975). The specificity of lowdose fenfluramine for the serotonin system is also suggested by studies in experimental animals showing serotonergic effects of fenfluramine at doses (less than 5 mg/kg) which do not alter levels of dopamine or HVA (Crunelli et al. 1980). To the authors' knowledge, attempts to block fenfluramine-mediated neuroendocrine responses in humans using 5-HT antagonists have not been undertaken, although clearly such studies would be of help in validating that the fenfluramine-induced prolactin rise in humans, like that found in experimental animals (Quattrone et al. 1978), is mediated by a serotonergic mechanism.

Other Factors of Possible Importance in Serotonin-Induced Prolactin Responses

Serotonin Receptors

Any attempt to interpret the effects of alterations in 5-HT activity must also take into account current knowledge suggesting that one mechanism by which 5-HT produces its postsynaptic effects is through interactions with membrane receptors. Thus another important potential determinant of serotonergic functions lies in the status of 5-HT receptor(s) (Samanin et al. 1980). At least in principle, the blunted prolactin rise observed here could also reflect a primary alteration in postsynaptic serotonin receptor sensitivity (i.e., "down regulation"), although direct evidence supporting this possibility is lacking at present. However, the possibility highlights the importance of ongoing research aimed at better defining the nature and regulation of 5-HT receptors. Several comprehensive reviews of this body of work have recently been published (Anderson 1983; Fillion 1983; Fuxe et al. 1983).

Current evidence suggests that several distinct populations of 5-HT receptors exist within the central nervous system of man and experimental animals. Evidence supporting this hypothesis is derived from in vitro radioligand binding studies in man (Cross 1982) and binding studies, amine release experiments, electrophysiological data, and behavioral studies in experimental animals (Middlemiss 1982). Binding data suggest the existence of at least two 5-HT recognition sites, the so-called 5-HT_1 site labeled by $^3\text{H-5-HT}$ and the 5-HT_2 site labeled by $^3\text{H-spiperone}$ (Peroutka and Snyder 1979, 1981; Peroutka et al. 1981). More recent evidence suggests that this classification may be an oversimplification, and that there are likely to be at least two distinct subtypes of 5-HT_1 receptors, the so-called 5-HT_{1A} and 5-HT_{1B} types (Pedigo et al. 1981). Biochemical data also exist suggesting that some 5-HT receptors are linked to cyclic adenosine monophosphate, and that these receptors may exist in both high- and low-affinity conformational states (Fillion 1983). Evidence from electrophysiological and behavioral studies is also consistent with the existence of multiple 5-HT receptors (Aghajanian 1981). Efforts to establish functional correlate for the 5-HT_2 recognition site suggest a significant correlation between the ability of a wide variety of antagonists to inhibit a number of 5-HT-mediated behavioral responses and their potency to inhibit radioligand binding to the $5\text{-}HT_2$ site (Leysen 1981; Peroutka and Snyder 1982; Leysen et al. 1982). The functional role of the $5\text{-}HT_1$ recognition site(s) remains unclear at present, and although attempts have been made to link this site with inhibitory autoreceptors (Martin and Sanders-Bush 1982; Gothert and Schlicker 1983), a 5-HT-sensitive adenylate cyclase (Fillion 1983; Peroutka et al. 1981), inhibition of dopamine release from the striatum (Ennis et al. 1981), a pure hyperlocomotion syndrome (Garner and Guy 1983), and basilar artery contraction (Peroutka et al. 1983), the evidence for each of these possibilities contains inconsistencies (Fozard 1983). The resolution of the important question of the function correlates of 5-HT recognition sites must await future research efforts.

Regarding the possibility that the blunted fenfluramine-induced prolactin response in depressives observed in this study reflects an alteration in 5-HT receptor responsivity, the next question would be which type of 5-HT receptor might be involved? As noted above, the principal mechanism by which fenfluramine is thought to alter serotonergic function is through release of presynaptic 5-HT. Thus the effects of fenfluramine are thought to be mediated primarily by endogenous 5-HT itself. On the basis of the fact that the affinity of 5-HT in ligand binding studies from data in experimental animals is 1000-fold greater for $5\text{-}HT_1$ receptors than for $5\text{-}HT_2$ receptors (Peroutka and Snyder 1982), one might reasonably hypothesize that the effects on prolactin secretion are mediated by $5\text{-}HT_1$ receptors. The recent development of more selective $5\text{-}HT_2$ antagonists (Janssen 1983) which cross the blood-brain barrier should allow testing of this hypothesis. The combined use of such drugs wich nonselective 5-HT antagonists, which possess $5\text{-}HT_1$ antagonist properties as well, might add further information in this regard. If in fact, as has been suggested by some authors (Pedigo et al. 1981), $5\text{-}HT_1$ receptors are heterogeneous, further definition of the subtype of $5\text{-}HT_1$ receptor must await the development of selective $5\text{-}HT_1$ antagonists and agonists.

Other Factors of Possible Importance to Serotonin-Induced Prolactin Responses

The interpretation of pharmacologically induced changes in prolactin concentrations must also take into account recent advances in our understanding of prolactin physiology. Evidence from studies in experimental animals suggests that 5-HT-mediated prolactin release is a complex function primarily involving stimulation of prolactin-releasing factor(s) (Clemens et al. 1978), but possibly also involving the inhibition of tuberoinfundibular dopamine (which functions as a prolactin-release-inhibiting factor) (Pilotte and Porter 1981) and possibly the stimulation of endogenous opioid activity (Somoza et al. 1983; Matsushita et al. 1981).

Thus other possibilities which in theory might also account for an altered prolactin response following enhancement of serotonergic function would include reduction in the function of either prolactin-releasing factor(s) or endogenous opioid activity on either a pre- or postsynaptic basis, although little evidence exists at present regarding either possibility. The importance of such factors as these must await future studies aimed at better defining the nature of prolactin-releasing factors and the importance of endogenous opioid activity in modulating prolactin release.

Another possiblity which might contribute to alteration in pharmacologically induced prolactin responsivity is a primary abnormality in pituitary lactotroph function. Although several other stimuli to prolactin secretion (TRH, methadone, morphine) have also been reported to be blunted in some (Gregoire et al. 1977; Linkowski et al. 1980; Extein et al. 1980; Judd et al. 1982; Ehrensing et al. 1974) but not all studies (Maeda et al. 1975) examining this issue in depressed patients, such studies cannot be used as evidence in this regard, since 5-HT may play a role in both opiate and TRH-induced prolactin changes. Specifically, the stimulating effects of morphine on prolactin have been reported to be reduced by the 5-HT synthesis inhibitor p-chlorophenylalanine (Dupont et al. 1981), and TRH-induced prolactin elevations have been shown to be inhibited by the serotonin antagonist metergoline (Ferrari et al. 1976). There is, in fact, increasing evidence that not only prolactin response differences but also hypothalamic-pituitary-adrenal axis abnormalities noted in some depressed patients may be reflections of more primary abnormalities in serotonergic and catacholaminergic pathways (Checkley 1980; Hatotani et al. 1981). In future studies it would be of interest to examine prolactin responses to several different stimuli in the same subject to evaluate how these different measures might relate to one another. Evaluation of the prolactin response to dopamine receptor antagonists might also be helpful in verifying the functional integrity of pituitary lactotrophs, in that evidence exists suggesting that the activity of the serotonergic system has no role in the activation of prolactin secretion induced by suppression of the inhibitory dopaminergic influence (Krulich et al. 1980).

Neuroendocrine Responses to Serotonin Precursors and Other Agents

Relatively few attempts have been made to use the pharmacologic challenge strategy in the assessment of serotonergic function in affectively ill patients. The report of impaired growth hormone (GH) release in response to insulin hypoglycemia among some depressives (Sachar et al. 1971), along with the reports that hypoglycemia-induced GH secretion could be inhibited by the serotonin antagonists cyproheptadine and methysergide (Bivens et al. 1973), gave rise to investigation of the GH response to the serotonin precursor 5-HTP in affectively ill subjects (Takahashi et al. 1973). These authors tested the GH response following L-5-HTP (200 mg orally) in 12 depressed patients and seven healthy subjects and found that 17 of 20 tests performed in the depressed patients showed blunted GH responses (i.e., GH increase of less than 5 ng/ml), compared to no abnormal tests out of seven in a group of healthy volunteers. While suggestive of a possible alteration in serotonergic function among depressives, other factors such as age and dopaminergic effects of L-5-HTP could not be excluded. More recently, Westenberg et al. (1982) studied neuroendocrine responses in 14 depressives and 12 healthy volunteers following both L-tryptophan (5.0 g orally) and L-5-HTP (200 mg orally). However, since administration of precursors failed to cause a significant change in hormone concentration in either group, the authors could only conclude that precursors at the doses used, administered orally, do not appear to provide a reliable index of serotonergic activity. Such results are not surprising in view of the suggestion from other precursor studies that oral administration of precursors may be less reliable in inducing hormone changes than parenteral administration (Charney et al. 1982; MacIndoe and Turkington 1973). Using intravenously administered tryptophan, Charney and co-workers have recently observed a blunted prolactin rise among some depressed subjects, consistent with the finding of this study and the hypothesis of a functional reduction of serotonergic activity in depression (G. Henninger, personal communication).

In contrast to the paucity of studies in affectively ill subjects, more information is available on the neuroendocrine effects of serotonin precursors in healthy subjects.

Neuroendocrine Effects of L-Tryptophan in Healthy Subjects

Prolactin Response

The effects of L-tryptophan on plasma prolactin have been studied by several groups. Both studies using intravenously administered L-tryptophan (5–10 g) have demonstrated a significant increase in plasma prolactin over baseline levels (Charney et al. 1982; MacIndoe and Turkington 1973). Studies examining the effects of oral tryptophan on plasma prolactin have shown more variable results. Studies using less than 90 mg/kg have failed to show a statistically significant increase in plasma prolactin (Fraser et al. 1979; Glass et al. 1980; Westenberg et al. 1982), whereas studies using 90–100 mg/kg have shown a trend toward an increase (Wiebe et al. 1977; Hyppa et al. 1979), and one study (Woolf and Lee 1977) using 150 mg tryptophan per kilogram demonstrated a significant increase at 90 min. These results suggest that doses of L-tryptophan exceeding 100 mg/kg may be required to increase plasma prolactin reliably and significantly in healthy subjects, and also suggest that the intravenous route of administration may be required to produce such changes reliably. Pretreatment with the 5-HT antagonist methysergide has been shown to reduce the effect of L-tryptophan on plasma prolactin secretion, supporting the hypothesis that the increase in plasma prolactin after tryptophan is mediated by a serotonergic mechanism (MacIndoe and Turkington 1973).

Growth Hormone Response

A significant increase in GH after acute tryptophan loading has been shown in most (Charney et al. 1982; Fraser et al. 1979; Hyppa et al. 1979; Glass 1979; Woof and Lee 1977; Koulu 1982) but not all studies (Westenberg et al. 1982; MacIndoe and Turkington). A dose-response relationship appears less clear than for plasma prolactin with significant increases in GH being noted after 2.0 g orally (Hyppa et al. 1979) and no consistent effect after doses of 10 g IV (MacIndoe and Turkington 1973). Considerable inter-subject variability in GH response has been noted in most studies. In both studies with placebo day comparisons, GH response after L-tryptophan was significantly greater than after placebo (Charney et al. 1982; Fraser et al. 1979). Furthermore, the increase in GH after L-tryptophan has been shown to be inhibited by pretreatment with either the serotonin antagonist cyproheptadine (Fraser et al. 1979) or melatonin (Koulu and Lammintausta 1979), supporting the hypothesis that the GH response is mediated by changes in serotonin activity.

Cortisol and Adrenocorticotropic Hormone Response

A definitive statement regarding the effect of L-tryptophan on plasma cortisol and adrenocorticotropic hormone (ACTH) levels cannot be made at present. The diurnal variation of these hormones makes comparison of L-tryptophan with placebo imperative, and to date such a placebo-controlled study addressing this question has not been reported. The results of existing studies of cortisol or ACTH response following L-tryptophan are contradictory, with evidence existing for an inhibitory effect (Woolf and Lee 1977), a stimulatory effect (Modlinger et al. 1979), and no effect (MacIndoe and Turkington 1973; Westenberg et al. 1982; Hyyppa et al. 1979) of L-tryptophan on cortisol or ACTH. Further appropriately designed studies of the effect of L-tryptophan on cortisol and ACTH are clearly warranted.

Luteinizing Hormone and Follicle-Stimulating Hormone Response

The effect of tryptophan administration on serum gonadotrophins has been studied by three groups, and none have been able to show a significant change in luteinizing hormone/follicle-stimulating hormone (LH/FSH) after either intravenous (10 g) or oral tryptophan administration (MacIndoe and Turkington 1973; Hyyppa et al. 1979; Woolf and Lee 1977).

Thyroid-Stimulating Hormone Response

Neither acute nor chronic oral administration of L-tryptophan (5–10 g/day) appears to affect either basal or TRH-stimulated thyroid-stimulating hormone (TSH) levels (Westenberg et al. 1982; Woolf and Lee 1977; Faber et al. 1977). Some preliminary evidence exists to suggest that acute intravenous administration of L-tryptophan (10 g) may reduce TSH (MacIndoe and Turkington 1973), but the lack of a placebo day comparison limits the interpretation of this result.

Neuroendocrine Effects of 5-Hydroxytryptophan in Healthy Subjects

Prolactin Response

The immediate serotonin precursor, L-5-HTP, administered intravenously to subjects pretreated with a decarboxylase inhibitor, has been shown to cause a significant increase in prolactin, an effect which is more pronounced in women than in men (Wirz-Justice et al. 1976; Puhringer et al. 1976; Lancranjan et al. 1977). After oral administration, L-5-HTP (200 mg) has also been reported to increase plasma prolactin in some studies (Kato et al. 1974; Yashimura et al. 1973) but not others (Handwerger et al. 1975; Beck-Peccoz et al. 1976; Westenberg et al. 1982).

Growth Hormone Response

Studies on the effects of orally administered 5-HTP (150–200 mg) on GH are mixed, with some groups reporting a significant increase in GH over baseline (Yoshimura et al. 1973; Imura et al. 1973; Nakai et al. 1974) but others unable to confirm this finding (Benkert et al. 1973; Muller et al. 1974; Westenberg et al. 1982). The only study including a full placebo day could show no significant difference between 5-HTP and placebo (Handwerger et al. 1975).

The effects of intravenously administered 5-HTP on GH have been studied by only one group. In subjects pretreated after 3 days with a decarboxylase inhibitor, infusion of 200 mg 5-HTP was found to cause a significant increase in GH over baseline, an increase that was also significantly greater than placebo infusion (Lancranjan et al. 1977; Puhringer et al. 1976). Support for the hypothesis that GH increases following 5-HTP are serotonin mediated derives from the demonstration that cyproheptadine, a nonselective 5-HT antagonist, blunts the GH rise after 5-HTP (Nakai et al. 1974).

Cortisol and Adrenocorticotropic Hormone Response

The effects of 5-HTP on ACTH and cortisol have not been well studied, and existing results are contradictory. Imura reported a significant increase above baseline in both ACTH and cortisol following 5-HTP (150 mg orally) in almost all subjects tested (Imura et al. 1973). In contrast, Westenberg was unable to confirm a significant change in cortisol (Westenberg et al. 1982). More recently, Meltzer and co-workers have reported significant increases in cortisol following 5-HTP (200 mg orally) (Meltzer et al. 1983). Clearly, more study is needed before a firm conclusion regarding the effects of 5-HTP on cortisol and ACTH can be made.

Luteinizing Hormone Response

The effects of 5-HTP on LH have not been extensively studied. In the only study of this subject, Benkert was unable to show a significant effect of D-L-5-HTP (150 mg orally) on episodic LH secretion in four male subjects (Benkert et al. 1973).

Thyroid-Stimulating Hormone Response

The effects of 5-HTP on TSH are not clear. Kato (1974) makes reference to unpublished results by that group showing that 5-HTP causes no significant change in

TSH. In contrast, Westenberg et al. (1982) reported a significant decrease in TSH both during baseline and treatment with 5-HTP, although a placebo day comparison was not performed. 5-HTP has also been reported to blunt the TSH response to TRH, but the 5-HT antagonist cyproheptadine has also been reported to have a similar effect on the TSH response (Paracchi et al. 1975; Egge et al. 1977).

Neuroendocrine Effects of Other Putative Serotonin Agonists

Parati et al. (1980) have reported that the 5-HT agonist quipazine, 50 mg orally, reliably increased serum cortisol in normal volunteers with no systematic effects on prolactin, GH, TSH, FSH, or LH. However, caution must be used in the interpretation of these results, in that quipazine produces significant gastrointestinal side effects and is thought to possess dopaminergic as well as serotonergic activity (Grabowska et al. 1974). Meltzer et al. (1983 b) have reported that N,N-dimethyltryptamine, an indole hallucinogen with 5-HT agonist properties, produces increases in prolactin, GH, and cortisol in experienced drug users, effects which are blocked in some subjects by pretreatment with cyproheptadine. In contrast, studies examining the neuroendocrine effects following acute administration of 5-HT reuptake inhibitors in man, including the newer selective agents such as citalopram and alaproclate, have failed to show significant hormonal changes (Hyttel 1982; Syvalahti et al. 1979) with the exception of studies with chlorimpramine, which has been reported to increase prolactin and GH (Laakman 1980).

Limitations of the Pharmacologic Challenge Strategy

Although many of the above data on the neuroendocrine effects of serotonergic agonists and the findings of our work with fenfluramine suggest the potential usefulness of the pharmacologic challenge strategy in assessing neurotransmitter function in humans, the use of neurotransmitter agonists or antagonists requires the recognition of several limitations of this strategy. First, the validity of the pharmacologic challenge approach depends in large part on the selectivity of the pharmacologic agent used to alter the functional activity of the transmitter system of interest (Checkley 1980). Particular attention must also be paid to the selectivity of metabolites of the parent compound. With regard to the use of 5-HT precursors as pharmacologic probes, this issue continues to receive attention, and the uses and limitations of transmitter precursor loading in general have recently been reviewed in detail (Curzon 1979). Several points warrant special mention. Regarding the use of tryptophan as a pharmacologic challenge agent, particular attention must be given to possible behavioral and neuroendocrine effects of enhanced trace amine synthesis (e.g., tryptamine), which may accompany high-dose tryptophan administration (Saavedra and Axelrod 1973). Furthermore, this effect may occur not only in 5-HT neurons but also in dopamine neurons (Curzon 1979). Evidence suggesting that extracerebral tryptamine may have relatively free access to the brain (Marsden and

Curzon 1978) also warrants consideration in the interpretation of studies using tryptophan to assess serotonergic function in man. Also, the possibility that some effects of tryptophan may be mediated via the kynurenine pathway (Lapin 1972; Handley and Miskin 1977) must also be entertained in the interpretation of studies using relatively high doses of tryptophan.

Regarding the use of 5-HTP as a neuroendocrine challenge agent, attention must be paid to the possibility that enhanced 5-HT synthesis may occur at nonphysiological sites as well as physiological sites, since the decarboxylase enzyme responsible for conversion of 5-HTP to 5-HT is widely distributed (Hokfelt et al. 1973).

Another important issue relevant to the use of the pharmacologic challenge strategy to assess central neurotransmitter function relates to the complex nature of the regulation of neuroendocrine responses. Although the use of peripheral hormone changes as response measures of altered neurotransmitter activity offers definite advantages in terms of safety, cost, accessibility, and reliability of measurement, the use of hormones as response measures has certain limitations as well. Perhaps the most limiting issue in this regard relates to the multiplicity of biological variables capable of influencing hormone levels, including such factors as age, sex, sexual maturity, stress, diet, other hormones, drugs, and diurnal rhythms, to name but a few. Great care should be taken to control for the influence of such variables. Increasing evidence also suggests multiple peptidergic influences on pituitary hormone release (Epelbaum et al. 1983), adding further complexity to the interpretation of drug-induced neuroendocrine changes. Furthermore, in that a given hormone may be influenced by several different major neurotransmitter systems which also interact with one another and with multiple peptidergic systems in a complex fashion, inferences drawn from the pharmacologic challenge strategy regarding the function of a particular neurotransmitter system must be made with caution.

In spite of these limitations, the results of neuroendocrine challenge studies remain important in that they evaluate a physiological consequence, albeit a complicated one, of a perturbation of neurotransmitter activity. Thus they complement more static measures of neurotransmitter function, such as measures of receptor number or levels of metabolites, which can provide no direct information on the net functional activity of the system.

Future Research Directions

The ability further to define serotonergic influences on the pathogenesis of psychiatric syndromes and on the mechanism of action of effective antidepressant treatments will depend in large part on the development of additional strategies for selectively altering serotonergic function. Important in this regard is recent evidence suggesting that neurotransmitters and neuropeptides may coexist in the same neuron and its terminals (Burnstock 1976). With regard to the serotonergic system, convincing evidence has emerged from work with experimental animals that the neuropeptides substance P and TRH coexist in many 5-HT neurons (Chan-Palay et al. 1978; Hokfelt et al. 1980a, b), and that such substances may be able to modify 5-HT receptor sensitivity (Fuxe et al. 1983). If such findings also prove to be true in man, then pharmacologic strategies targeting cotransmitter receptors may be worth investigating as an additional method of altering serotonergic function. An emphasis in future work should also be placed on the development and investigation of more selective serotonin agonists and antagonists and the study of the behavioral effects of these drugs on affective symptomatology. Along similar lines, studies of the effects of somatic treatments (tricyclic and MAO-inhibiting antidepressant drugs, electroconvulsive therapy, sleep deprivation) on 5-HT-agonist-mediated responses may also prove useful in furthering our limited understanding of the relationship between serotonin and behavior in man.

References

Abrams R (1978) Serotonin and affective disorders. In: Essman WB (ed) Serotonin in health and disease, vol III. The central nervous system. Spectrum, Jamaica, NY

Aghajanian GK (1981) Modulatory role of serotonin at multiple receptors in brain. In Jacobs B, Gelperin A (eds) Serotonin neurotransmission and behavior. MIT Press, Cambridge

Aloi J, Insel T, Mueller EA, Murphy DL (1984) Neuroendocrine and behavioral effects of *m*-chlorophenylpiperazine administration in rhesus monkeys. Life Sci 34:1325–1331

Anderson JL (1983) Serotonin receptor changes after chronic antidepressant treatments: ligand binding, electrophysiological and behavioral studies. Life Sci 32:1791

Aprison MH, Takahashi R, Tachiki K (1978) Hypersensitive serotonin receptors involved in clinical depression – a theory. In: Haber B, Aprison MH (eds) Neuropharmacology and behavior. Plenum, New York

Beck-Peccoz P, Ferrari C, Rondena M, Paracchi H, Faglia G (1976) Failure of oral 5-hydroxytryptophan administration to effect prolactin secretion in man. Horm Res 7:303

Benkert O, Laakmann G, Souvatzoglou A, Von Werder K (1973) Missing indicator function of growth hormone and luteinizing hormone blood levels for serotonin and dopamine concentration in the human brain. Neural Transm 34:291

Bivens CH, Lebovitz HE, Feldman J (1973) Inhibition of hypoglycemia induced growth hormone secretion by the serotonin antagonists cyproheptadine and methysergide. N Engl J Med 289:236

Briley MS, Langer SZ, Raisman R (1980) Tritiated imipramine binding sites are decreased in platelets of untreated depressed patients. Science 209:303

Bunney WE (1983) Biological markers. Psychiatr Ann 13 (5):366

Bunney WE, Post RM, Anderson AE, Kopanda RT (1977) A neuronal receptor sensitivity mechanism in affective illness (a review of evidence). Commun Psychopharmacol 1:395

Burnstock G (1976) Do some nerve cells release more than one transmitter? Neuroscience 1:239

Caufriez A, Desir D, Szyper M, Robyn C, Copinschi G (1981) Prolactin secretion in Cushing's disease. J Clin Endocrinol Metabol 53:843

Chan-Palay V, Johansson G, Palay SL (1978) Serotonin and substance P coexist in neurons of the rat's central nervous system. Proc Natl Acad Sci USA 75:1582

Charney DS, Heninger GR et al. (1982) The effect of intravenous L-tryptophan on prolactin and growth hormone and mood in healthy subjects. Psychopharmacology (Berlin) 77:217

Chase TN, Shoulson I, Carter AC (1976) Serotonergic functions in man. Monogr Neural Sci 3:8

Checkley SA (1980) Neuroendocrine tests to monoamine function in man: a review of basic theory and its application to the study of depressive illness. Psychol Med 10:35

Clemens JA, Roush ME, Fuller RW (1978) Evidence that serotonin neurons stimulate secretion of prolactin releasing factor. Life Sci 22:2209

Cleinschmidt BV, McGuffin JC (1978) Pharmacological differentiation of the central 5-hydroxytryptamine-like actions of MK-212, *p*-methocyamphetamine and fenfluramine in an in vivo model system. Eur J Pharmacol 50:369

Coppen A, Shaw DM, Farrell JP (1963) Potentiation of the antidepressive effects of a monoamine oxidase inhibitory by tryptophan. Lancet 1:79

Cross AJ (1982) Interactions of ^{3}H-LSD with serotonin receptors in human brain. Eur J Pharmacol 82:77

Crunelli V, Bernasconi S, Samanin R (1980) Effects of d- and l-fenfluramine on striatal HVA concentrations in rats after pharmacological manipulation of brain serotonin. Pharmacol Res Commun 12 (3):215

Curzon G (1979) Transmitter precursor loading-uses and limitations. In: Saletu B et al. (ed) Neuropsychopharmacology. Pergamon, New York

Dupont A, Barden N, Labrie F, Ferland L, Cusan L (1981) Opiates and neuroendocrine regulations. In: Hrdina PD, Singhal RL (eds) Neuroendocrine regulation and altered behavior. Croom Helm, London

Ehrensing RH, Kastin AJ, Schach DS (1974) Affective state and prolactin responses after repeated injections of thyrotropin-releasing hormone in depressed patients. Am J Psychiatry 131:714

Egge AC, Rogol AD, Varma MM, Blizzard RM (1977) Effect of cyproheptadine on TRH-stimulated PRL and TSH release in man. J Clin Endocrinol Metabol 44:210

Ennis C, Kemp JD, Cox B (1981) Characterization of inhibitory 5-hydroxytryptamine receptors that modulate dopamine release in the striatum. J Neurochem 36:1515

Epelbaum J, Enjalbert A, Drouna S, Kordon C (1983) Involvement of newly discovered neuropeptides in the control of neuroendocrine process. In: Endroczi E et al. (ed) Integrative neurohumoral mechanisms. Elsevier, Amsterdam

Extein I, Pottash ALC, Gold MS (1980) Deficient prolactin response to morphine in depressed patients. Am J Psychiatry 137:845

Faber J, Hagen C, Kirkegaard C, Lauridsen UB, Moller SE (1977) Lack of effects of L-tryptophan on basal and TRH-stimulated TSH and prolactin levels. Psychoneuroendocrinology 2:413

Ferrari C, Paracchi A, Rondena M, Beck-Peccoz P, Faglia G (1976) Effect of two serotonin antagonists on prolactin and thyrotropin secretion in man. Clin Endocrinol 5:575

Fillion G (1983) 5-Hydroxytryptamine receptors in brain. In: Iversen LL, Iversen SD, Snyder SH (eds) Handbook of psychopharmacology, vol 17. Plenum, New York

Fozard JR (1983) Functional correlates of 5HT-1 recognition sites. Trends Pharmacol Sci 4 (7):288–290

Fraser W, Tucker St H, Grubb SR, Wigand JP, Blackard WG (1979) Effect of L-tryptophan on growth hormone and prolactin release in normal volunteers and patients with secretory pituitary tumors. Hor Metab Res 11:149

Fuller RW, Perry KW, Clemens JA (1976) Elevation of 3,4-dihydroxyphenylacetic acid concentrations in rat brain and stimulation of prolactin secretion by fenfluramine: evidence of antagonism at dopamine receptor sites. J Pharm Pharmacol 28:643

Fuller RW, Snoddy HD, Clemens JA, Molloy BB (1982) Effect of norfenfluramine and two structural analogous on brain 5-hydroxyindoles and serum prolactin in rats. J Pharm Pharmacol 34:449

Fuxe K, Ogren SO, Agnati LF et al. (1983) Chronic antidepressant treatment and central 5HT synapses. Neuropharmacology 22:389

Garattini S, Buczko W, Jori A, Samanin R (1975) The mechanism of action of fenfluramine. Postgrad Med J 51 (1):27

Garattini S, Caccia S, Mennini T, Samanin R, Consolo S, Landinsky H (1979) Biochemical pharmacology of the anorectic drug fenfluramine: a review. Curr Med Res Opin 6 (1):15

Garner CR, Guy AP (1983) Behavioral effects of RU 24969, a 5HT-1 receptor agonist, in the mouse. Br J Pharmacol 78:96 P.

Glass AR, Schaaf M, Dimond RC (1979) Absent growth hormone response to L-tryptophan in acromegaly. J Clin Endocrinol Metabol 48:664

Glass AR, Smallridge RC, Schaff M, Dimond RC (1980) Absent prolactin response to L-tryptophan in normal and acromegalic subjects. Psychoneuroendocrinology 5:261

Gothert M, Schlicker E (1983) Autoreceptor mediated inhibition of ^{3}H-5HT release from rat brain cortex slices by analogues of 5-hydroxytryptamine. Life Sci 32:1183

Grabowska M, Antklewicz L, Michaluk J (1974) A possible interaction of quipazine with central dopamine structures. J Pharm Pharmacol 26:74

Gregiore E, Brauman H, DeBuch H (1977) Hormone release in depressed patients before and after recovery. Psychoneuroendocrinology 2:303

Handley SL, Miskin RC (1977) The interaction of some kynurenine pathway metabolites with 5-hydroxytryptophan and 5-hydroxytryptamine. Psychopharmacology (Berlin) 51:305

Handwerger S, Plonk JW, Lebovitz JW, Bivens CH, Feldman JM (1975) Failure of 5-hydroxytryptophan to stimulate prolactin and GH secretion in man. Horm Metabol Res 7:214

Hatotani N, Nomura J, Kitayama I (1981) Neuroendocrine studies on the pathogenesis of depression. In: Hrdina PD, Singhal RL (eds) Neuroendocrine regulation and altered behavior. Croom Helm, London

Hokfelt T, Fuxe K, Goldstein M (1973) Immunohistochemical localisation of aromatic L-amino acid decarboxylase (DOPA decarboxylase) in central dopamine and 5-hydroxytryptamine nerve cell bodies of the rat. Brain Res 53:175

Hokfelt T, Johansson O, Ljungdahl A et al. (1980a) Peptidergic neurons. Nature 284:515

Hokfelt T, Lundberg JM, Schultzberg M, Ljungdahl A, Rehfeld J (1980b) Coexistence of peptides and putative transmitters in neurons. Adv Biochem Psychopharmacol 22:1

Hyttel J (1982) Citalopram: pharmacological profile of a specific serotonin uptake inhibitor with antidepressant activity. Prog Neuropharmacol Biol Psychiatry 6:277

Hyyppa MT, Kytomaki O, Rautakorpi I, Syvalahti E (1975) Effects of tryptophan loading on neuroendocrine regulation in man. Acta Endocrinol [Supple] Copenh 199:315

Hyyppa MT, Tapani J, Liira J, Langvik V-A, Kytomaki O (1979) L-Tryptophan treatment and the episodic secretion of pituitary hormones and cortisol. Psychoneuroendocrinology 4:29

Imura H, Nakai Y, Yoshimi T (1973) Effect of 5-hydroxytryptophan (5HTP) on growth hormone and ACTH release in man. J Clin Endocrinol Metab 36:204

Janssen PA (1983) 5HT-2 receptor blockade to study serotonin-induced pathology. Trends Pharmacol Sci 4 (5):198

Judd LL, Risch SC, Parker DC, Janowsky DS, Segal L, Huey Y (1982) Blunted prolactin response: a neuroendocrine abnormality manifested by depressed patients. Arch Gen Psychiatry 39:1413

Kannengiesser MH, Hunt PF, Raynaud JP (1976) Comparative action of fenfluramine on the uptake and release of serotonin and dopamine. Eur J Pharmacol 35:35

Kato Y, Nakai Y, Imura H, Chihara K, Ohgo S (1974) Effect of 5-hydroxytryptophan (5HTP) on plasma prolactin levels in man. J Clin Endocrinol Metab 38:695

Koulu MO (1982) Re-evaluation of L-tryptophan-stimulated human growth hormone secretion: a dose related study with a comparison with L-dopa and apomorphine tests. J Neural Transm 55:269

Koulu M, Lammintausta R (1979) Effect of melatonin on L-tryptophan- and apomorphine-stimulated growth hormone secretion in man. J Clin Endocrinol Metab 49:70

Krulich L, Coppings RJ, Giachetti A, McCann SM, Mayfield A (1980) Lack of evidence that central serotonergic system plays a role in the activation of prolactin secretion following inhibition of DA synthesis or blockade of DA receptors in male rat. Neuroendocrinology 30:133

Kuhn CM, Vogel RA, Mailman RB, Mueller RA, Schanberg SM, Breese GR (1981) Effect of 5,7 dihydroxytryptamine on serotonergic control of prolactin secretion and behavior in rats. Psychopharmacology (Berlin) 73:188

Laakmann G (1980) Effect of antidepressants on the secretion of pituitary hormones in healthy subjects, neurotic depressive patients and endogenous depressive patients. Nervenarzt 51(12):725

Lancranjan I, Wirz-Justice A, Puhringer W, Del Polzo E (1977) Effect of L-5-hydroxytryptophan on growth hormone and prolactin secretion in man. J Clin Endocrinol Metabol 45:588

Lapin IP (1972) Interaction of kynurenine and its metabolites with tryptamine, serotonin and its precursors and oxyremorine. Psychopharmacology (Berlin) 26:237

Lapin IP, Oxenkrug GF (1969) Intensification of the central serotonergic process as a possible determinant of the thymoleptic effect. Lancet 1:132

Leysen JE (1981) Serotonergic receptors in brain tissue: properties and identification of various ^{3}H-ligand binding sites in vitro. J Physiol (Paris) 77:351

Leysen JE, Niemegeers CJE, Van Neuten JM, Laduron P (1982) [^{3}H-] Ketanserin (R41 468), a selective ^{3}H-ligand for serotonin-2 receptor binding sites. Mol Pharmacol 21:301

Linkowski P, Brauman H, Mendlewicz J (1980) Prolactin secretion in women with unipolar and bipolar depression. Psychiatry Res 3:265

MacIndoe JH, Turkington RW (1973) Stimulation of human prolactin secretion by intravenous infusion of L-tryptophan. J Clin Invest 52:1972

Maeda K, Kato Y, Ohgo S et al. (1975) Growth hormone and prolactin release after injection of thyrotropin releasing hormone in patients with depression. J Clin Endocrinol Metabol 40:501

Marsden CA, Curzon G (1978) The contribution of tryptamine to the behavioral effects of L-tryptophan in tranylcypromine treated rats. Psychopharmacology (Berlin) 57:71

Martin LL, Sanders-Bush E (1982) Comparison of the pharmacological characteristics of 5HT-1 and 5HT-2 binding sites with those of serotonin autoreceptors which modulate serotonin release. Naunyn Schmiedebergs Arch Pharmacol 321:165

Martin JB, Reichlin S, Brown GM (1977) Hypothalamic control of anterior pituitary secretion. In: Martin JB, Reichin S, Brown GM (eds) Clinical neuroendocrinology. Davis, Philadelphia

Matsushita N, Kato Y, Katakami H, Shimatsu A, Imura H (1981) Inhibition by naloxone of prolactin release induced by L-5HTP in rats. Proc Soc Exp Biol Med 168:282

Meltzer H, Wiita B, Tricon BJ, Simonovic M, Fang V, Manov G (1982) Effect of serotonin precursors and serotonin antagonists on plasma hormone levels. In: Ho BT (ed) Biological psychiatry. Raven, New York

Meltzer HY, Umberkoman-Wiita, B, Robertson A, Tricon BJ, Lowy M (1983a) Enhanced serum cortisol response to 5-hydroxytryptophan in depression and mania. Life Sci 33:2541–2549

Meltzer HY, Boutros NN, Simonovic M, Gudelsky GA, Fang VA (1983b) Hallucinogenic drugs and neuroendocrine secretion. In Erdroczi E et al. (eds) Integrative neurohumoral mechanisms. Elsevier, New York

Middlemiss DN (1982) Multiple 5-hydroxytryptamine receptors in the central nervous system of the rat. In: DeBelleroche J (ed) Presynaptic receptors: mechanism and function. Ellis Horwood, Chichester

Modlinger RS, Schonmuller JM, Arora SP (1979) Stimulation of aldoesterone, renin and cortisol by tryptophan. J Clin Endocrinol Metabol 48:599

Muller EE, Brambilla F, Cavagnini F, Peracchi M, Panerai A (1974) Slight effect of L-tryptophan on growth hormone release in normal human subjects. J Clin Endocrinol Metabol 39:1

Murphy DL, Campbell IC, Costa JL (1978) The brain serotonin system in the affective disorders. Prog Neuropsychopharmacol 2:1

Nakai Y, Imura H, Sakurai H, Kurahachi H, Yoshimi T (1974) Effect of cyproheptadine on human growth hormone secretion. J Clin Endocrinol Metabol 38:446

Ogren SO, Fuxe K, Agnan LF et al. (1979) Re-evaluation of the indoleamine hypothesis of depression: evidence for a reduction of functional activity of central 5HT systems by antidepressants. J Neural Transm 46:85

Olsen RW, Reisine T, Yamamura HI (1980) Neurotransmitter receptors – biochemistry and alterations in neuropsychiatric disorders. Life Sci 27:801

Paracchi A, Ferrari C, Faglia G (1975) Catecholaminergic and serotonergic influences on thyrotropin secretion in man. Acta Endocrinol [Supple] (Copenh) 80:199

Parati EA, Zanardi P, Cochi D, Caraceni T, Muller EE (1980) Neuroendocrine effects of quipazine in man in health state or with neurological disorders. J Neural Transm 47:273

Paul SM, Rehavi M, Skolnick P, Ballenger JC, Goodwin FK (1981) Depressed patients have decreased binding of ^{3}H-imipramine to platelet serotonin transporter. Arch Gen Psychiatry 38:1315

Pedigo NW, Yamamura HI, Nelson DL (1981) Discrimination of multiple ^{3}H-5-hydroxytryptamine binding sites by the neuroleptic spiperone in rat brain. J Neurochem 36:220

Perez-Lopez FR, Robyn C (1974) Studies on human prolactin physiology. Life Sci 15:599

Perez-Lopez FR, Gomez-Agudo G, Avos MD (1981) Serum prolactin and thyrotrophin responses to thyrotrophin-releasing hormone at different times of the day in normal women. Acta Endocrinol 97:7

Peroutka SJ, Snyder SH (1979) Multiple serotonin receptors: differential binding of ^{3}(H)5-hydroxytryptamine, ^{3}(H)lysergic acid diethylamide and ^{3}(H) spiroperidol. Mol Pharmacol 16:687

Peroutka SJ, Snyder SH (1981) Two distinct serotonin receptors: regional variations in receptor binding in mammalian brain. Brain Res 208:339

Peroutka SJ, Snyder SH (1982) Radioactive ligand binding studies: identification of multiple serotonin receptors. In: Osborne NN (ed) Biology of serotonergic transmission. Wiley, New York

Peroutka SJ, Lebovitz RM, Snyder SH (1981) Two distinct central serotonin receptors with different physiologic functions. Science 212:828

Peroutka SJ, Noguchi M, Tolner D, Allen G (1983) Serotonin mediated contraction of canine basilar artery: mediation by 5HT-1 receptors. Brain Res 259:327

Pilotte NS, Porter JC (1981) Dopamine in hypophysial portal plasma and prolactin in systemic plasma of rats treated with 5-hydroxytryptamine. Endocrinology 108:2137

Pinder RM, Brogden RN, Sawyer PR, Speight TM, and Avery GS (1975) Fenfluramine: a review of its pharmacological properties and therapeutic efficacy in obesity. Drugs 10:241

Preziosi P (1983) Serotonin control of prolactin release: an intriguing puzzle. Trends Pharmacol Sci 4:171

Puhringer W, Wirz-Justice A, Lancranjan I (1976) Mood elevation and pituitary stimulation after I.V. L-5-HTP in normal subjects: evidence for a common serotonergic mechanism. Neurosci Lett 2:349

Quattrone A, diRenzo G, Schettini G, Teceschi G, Scopacasa F (1978) Increased plasma prolactin levels induced in rats by d-fenfluramine: relation to central serotonergic stimulation. Eur J Pharmacol 49:163

Saavedra JM, Axelrod J (1973) Effect of drugs on the tryptamine content of rat tissue. J Pharmacol Exp Ther 185:523

Sachar EJ, Finkelstein J, Hellman L (1971) Growth hormone response in depressive illness. I. Response to insulin tolerance tests. Arch Gen Psychiatry 25:263

Samanin R, Mennini T, Ferraris A, Bendotti C, Borsini F (1980) Hyper- and hyposensitivity and central serotonin receptors: ^{3}H-serotonin binding and functional studies in the rat. Brain Res 189:449

Shoulson I, Chase TN (1975) Fenfluramine in man: hypophagia associated with diminished serotonin turnover. Clin Pharmacol Ther 8:616

Siever LJ, Murphy DL, Slater S, DelaVega E, Lipper S (1983) Plasma prolactin changes following fenfluramine in depressed patients compared to controls: an evaluation of central serotonergic responsivity in depression. (in press)

Slater S, DelaVega E, Skyler J, Murphy DL (1976) Plasma prolactin stimulation by fenfluramine and amphetamine. Psychopharmacol Bull 123:26

Somoza GM, Larrea GA, Becu D, Cardinali DP, Libertun C (1983) Inhibition by naloxone of the serotonin-induced prolactin release in free moving rats. J Neural Transm 56:97

Stahl SM, Woo DJ, Metford IN, Berger PA, Ciaranello RD (1983) Hyperserotonemia and platelet serotonin uptake and release in schizophrenia and affective disorders. Am J Psychiatry 14:26

Stanley M, Mann JJ (1983) Increased serotonin-2 binding sites in frontal cortex of suicide victims. Lancet 1:214

Stanley M, Virgilio J, Gershon S (1982) Tritiated imipramine binding sites are decreased in the frontal coretex of suicides. Science 216:1337

Sylvalahti E, Eneroth P, Ross SB (1979) Acute effects of zimelidine and alaproclate, who inhibitors of serotonin uptake, on neuroendocrine function. Psychiatry Res 1:111

Takahashi S, Kondo H, Yoshimura M, Ochi Y, Yoshimi T (1973) Growth hormone responses to administration of L-5-hydroxytryptophan (L-5HTP) in manic-depressive psychoses. Folia Psychiatr Neurol Jpn 27 (3):197

Trulson ME, Jacobs BL (1976) Behavioral evidence for the rapid release of CNS serotonin by PCA and fenfluramine. Eur J Pharmacol 36:149

Usdin E, Hanin I (eds) (1982) Biological markers in psychiatry and neurology. Pergamon, New York

Van Praag HM (1962) A critical investigation of the significance of MAO inhibition as a therapeutic principle in the treatment of depressions. Ph D thesis, University of Utrecht

Van Praag HM (1984) Central monoamines and the pathogenesis of depression. In: Van Praag HM, Lader, Rafaelsen, Sachar (eds) Handbook of biological psychiatry, vol 14. Marcel Dekker, New York

Westenberg HGM, Van Praag HM, DeJong J, Thijssen JHH (1982) Postsynaptic serotonergic activity in depressive patients: evaluation of the neuroendocrine strategy. Psychiatry Res 7:361

Wiebe RH, Handwerger S, Hammond CB (1977) Failure of L-tryptophan to stimulate prolactin secretion in man. J Clin Endocrinol Metabol 45:1310

Wirz-Justice A, Puhringer W, Lacoste P, Graw P, Gastpar M (1976) Intravenous L-5HTP in normal subjects: an interdisciplinary precursor loading study. III. Neuroendocrinological and biochemical changes. Pharmacopsychiatria 9:277

Woolf PD, Lee L (1977) Effect of the serotonin precursor, tryptophan on pituitary hormone secretion. J Clin Endocrinol Metabol 45:123

Yoshimura M, Ochi Y, Miyataki T, Shiomi K, Hachiya T (1973) Effect of L-5HTP on the release of growh hormone, TSH and insulin. Endocrinol Jpn 20:135

Cerebrospinal Fluid Amine Metabolite Studies in Depression: Research Update

D. C. JIMERSON and W. BERRETTINI

Introduction

Hypotheses implicating alterations of catecholamine and indoleamine neurotransmitters in affective illness arose from observations on the mood and behavioral effects of amine-altering drugs in laboratory animals and man. Further evidence came from studies showing that antidepressant drugs had major effects on the synaptic function of the amine neurotransmitters, including the sensitivity of pre- and postsynaptic receptors (Schildkraut 1965; Bunney and Davis 1965; Garver and Davis 1979; Van Praag 1977; Charney et al. 1981).

Clinical studies of these central nervous system (CNS) neurotransmitters have rested heavily on measurement of the levels of their major metabolites in cerebrospinal fluid (CSF). Studies in laboratory animals showed that the major metabolites of the catecholamines and serotonin provide a useful index of turnover of the parent neurotransmitter. Because of the intimate relationship of CSF with brain extracellular fluid, and because of its relative isolation from the periphery by the blood-brain barrier, CSF provided a window for study of central metabolites in man. Indeed, pharmacologic alterations of catecholamine and serotonin function produced the expected changes in CSF metabolite levels (Goodwin et al. 1973).

In spite of the initial optimism about these studies, reports on amine metabolite levels began to yield mixed results (reviewed in Goodwin et al. 1975; Jimerson et al. 1976; Post et al. 1980), leading investigators to reexamine methodological problems. Table 1 illustrates the lengthening list of methodological variables to be controlled in clinical CSF studies.

Table 1. Methodological variables to be controlled in CSF metabolite studies

1. Diagnostic criteria	11. CSF aliquot
2. Phase of illness	12. CSF storage conditions
3. Control group	13. Assay methodology
4. Medication status	14. Sex effects
5. Other physical illness	15. Age effects
6. Diet	16. Height effects
7. Motor activity	17. Body weight
8. Time of day	18. Probenecid level
9. LP position	19. State anxiety
10. Site of LP	20. Personality variables

LP, lumbar puncture

The present chapter reviews the results of recent CSF studies of catecholamine and indoleamine metabolites in depression. Included is a discussion of current methodological issues, such as the influence of plasma levels of the lipophilic norepinephrine metabolite MHPG (3-methoxy-4-hydroxyphenylglycol) on CSF concentrations, and clinical issues, such as the relationship of metabolite levels to specific symptoms as opposed to discrete diagnostic syndromes. Because of space limitations, we have not reviewed the numerous neurochemical processes that influence CSF metabolite levels, including not only rates of neurotransmitter release, but also rates of transmitter reuptake, transmitter metabolism, and metabolite transport into and out of CSF. Moreover, any interpretation of the behavioral correlates of neurotransmitter turnover needs to include consideration of alterations in receptor sensitivity, postsynaptic receptor output (e.g., second messenger systems), and the probable influence of cotransmitters and other neuromodulators.

Norepinephrine Metabolites in Depression

Evidence that the major metabolite of brain norepinephrine (NE) is MHPG and evidence that MHPG levels in brain reflect alterations of NE turnover (Meek and Neff 1973; Chase et al. 1973) prompted studies of CSF MHPG in patients with major depression. In general, these studies have not supported a simple hypothesis of decreased central NE turnover in this patient group (Table 2), and recent studies actually suggest that some depressed women may have increased MHPG levels (Koslow

Table 2. CSF 3-methoxy-4-hydroxyphenylglycol (MHPG) levels in depressed patients and controls

Author	Assay method	Med-free interval (days)	Depressed[a] patients	Controls
Wilk et al. 1972	GC	7	17.0±2 (5)	15 ±1 (19)
Shaw et al. 1973	GC	–[e]	11.9±0.6 (22)	10.8±0.8 (13)
Post et al. 1973	GC	14	8.9±1.1 (25)[c]	16.3±2.1 (10)
Shopsin et al. 1974	GC	7	15.8 (11–20) (8)	15.9 (10–23) (18)
Subrahmanyam 1975	GC	2	14.2±2.6 (24)[c]	20.6±2.4 (12)
Ashcroft et al. 1975	GC	–	12.0±2.4 (7)	13.0±2.5 (11)
Vestergaard et al. 1978	GC	5	12.0±0.8 (27)	10.4±0.9 (21)
Berger et al. 1980	GC/MS	14	9.9±0.8 (11)	9.2±0.6 (22)
Ågren 1980a	GC/MS	7	10.6±0.3 (33)	11.2±2.8[d]
Oreland et al. 1981	GC/MS	7	9.4±0.4 (18)	9.1±0.3 (42)
Träskman et al. 1981	GC/MS	2–60	9.8±0.9 (7)	9.7±0.3 (45)
Jimerson et al. 1984a	GC/MS	14	10.5±1.0 (12)[b]	8.1±0.4 (11)
Koslow et al. 1983	GC/MS	11	8.8±0.2 (99)[b]	8.0±0.2 (61)

GC, gas chromatography; MS, mass spectrometry
[a] CSF MHPG level in ng/ml±SEM (no. of subjects)
[b] Patients significantly higher than controls
[c] Patients significantly lower than controls
[d] SD
[e] Depressed patients studied prior to treatment

et al. 1983; Jimerson et al. 1984a). Measurement of CSF levels of NE itself have also yielded mixed results (Post et al. 1978a; Christensen et al. 1980), with data by Post et al. (1980) showing elevated NE levels in comparison to healthy volunteers but not in comparison to a carefully selected group of neurological controls. Some discrepancies in the earlier studies in Table 2 may have resulted from the limited sensitivity and specificity of the gas chromatographic analytic methods. Difficulty of obtaining healthy controls may also have influenced results, for example, in the study by Post et al. (1973), where the parents of hospitalized children may have had elevated MHPG levels in relationship to stressful family events. The recent study by Charney et al. (1982) showing progressive elevation of plasma MHPG during the 2 weeks following antidepressant drug withdrawal raises questions about the adequacy of the 2-week drug-free interval prior to lumbar puncture (LP) in even the most recent studies. Since a number of these patients may actually have had longer drug-free intervals, future reanalysis of the data may clarify the antidepressant withdrawal question.

Another potentially confounding influence on CSF levels of MHPG has emerged from recent evidence that plasma levels of the metabolite can affect the CSF concentration (Kopin et al. 1983). Since free MHPG is a relatively small lipophilic molecule, it diffuses as rapidly as water out of nervous tissue (Kessler et al. 1976) and across biological membranes such as the blood-brain and blood-CSF barriers. This diffusion may account in large part for the correlation between plasma and CSF levels of MHPG in healthy volunteers (Jimerson et al. 1981) and subhuman primates (Elsworth et al. 1982). Thus reduction of peripheral sympathetic activity, leading to reduced plasma MHPG levels, could result in turn in a reduction in CSF MHPG levels; such a reduction might erroneously be interpreted as reflecting decreased CNS NE turnover if the peripheral changes were ignored. An example of this situation has been observed in patients with idiopathic orthostatic hypotension (Polinsky et al. 1984).

There is evidence that peripheral sympathetic nervous system activity may be decreased (Schildkraut 1978) or increased (Lake et al. 1982; Koslow et al. 1983) in some subgroups of depressed patients. Thus simultaneous assessment of plasma MHPG levels at the time of LP is needed to assure that apparent patient differences in CSF MHPG level do not solely reflect variations in plasma level. For comparing CNS MHPG production across patient groups, one can use values of Mc/k_2, calculated as $[C]-k_1/k_2[P]$, where Mc is the rate of CNS production of MHPG, k_1 and k_2 are rate constants (due mainly to diffusion) for entry of MHPG from plasma to CSF and from CSF to plasma respectively, and $[C]$ and $[P]$ are the CSF and plasma concentrations of MHPG. For clinical studies, k_1/k_2 has been estimated as 0.9 (Kopin et al. 1983). In a recent preliminary study of simultaneous CSF and plasma MHPG levels in 12 depressed patients, both the CSF MHPG level and M_c/k_2 index were significantly elevated in comparison to values for 11 age-matched controls (Jimerson et al. 1984a) (Fig. 1). Further studies are necessary, however, to rule out the possibility that variations in plasma MHPG levels have contributed to the varied results of previous CSF MHPG studies in depression.

Recent evidence also suggests that MHPG levels in patient studies could be influenced not only by the diagnostic syndrome of depression, per se, but also as correlates of underlying personality traits or state-related symptoms at the time of the

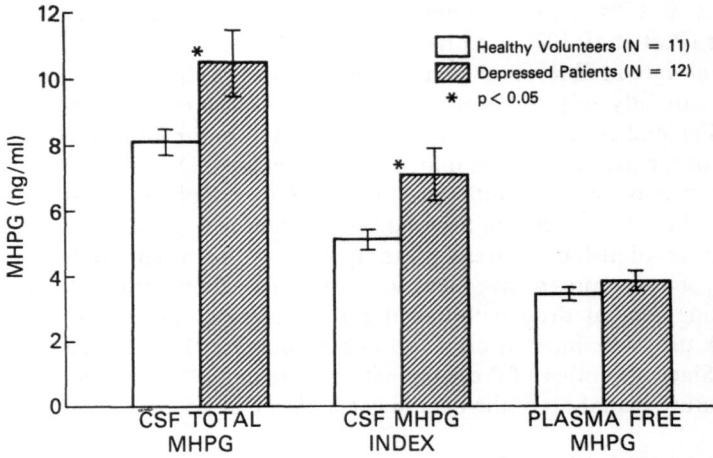

Fig. 1. CSF MHPG concentration was significantly higher in medication-free depressed patients than in healthy controls. Since the plasma MHPG level was not significantly different for the two groups, the CNS metabolite contribution (CSF MHPG index, Mc/k_2) was significantly elevated in the depressed patients (Jimerson et al. 1984a)

LP. Thus plasma MHPG levels in healthy volunteers were correlated with Minnesota Multiphasic Personality Inventory (MMPI) personality variables of depression, hysteria, and hypochondriasis (Post et al. to be published). CSF MHPG levels in healthy controls and CSF NE levels in depressed patients were related to state anxiety at the time of the LP (Ballenger et al. 1984; Post et al. 1978a). Because of the relationship between anxiety and increased MHPG levels, the use of anxiolytic medications during the "drug-free" period prior to LP could indirectly affect CSF MHPG measured. In line with earlier observations that CSF MHPG was not generally correlated with severity of depression, Ågren (1980a) has shown that in unipolar depressed patients, symptoms of lack of energy, overt anger, self-pity, and somatic preoccupation were related to high CSF MHPG; and discouragement, demandingness, subjective anger, and suicidal tendencies were related to decreased CSF MHPG. Future studies may benefit from covarying out the influences of state and trait variables on CSF metabolites prior to comparing patients and controls, perhaps revealing an underlying association between the syndrome of major depression and altered metabolite levels.

Dopamine Metabolites in Depression

Evidence for altered dopamine (DA) function in depressed patients initially derived from observations on mood changes associated with drugs with prominent DA effects (Murphy et al. 1971; Randrup et al. 1975). Subsequent laboratory studies have shown that antidepressant drugs block DA reuptake (Halaris et al. 1975; Randrup and Braestrup 1977) and down-regulate presynaptic DA receptors (Serra et al. 1979;

Antelman and Chiodo 1981; Lee and Tang 1982), effects which would result in increased neuronal release of DA. Moreover, relatively selective DA receptor agonists such as piribedil and bromocriptine have significant antidepressant effects (Post et al. 1978 b; Shopsin and Gershon 1978; Colonna et al. 1979; Waehrens and Gerlach 1981; Nordin et al. 1981). Neuroendocrine studies have not, however, revealed specific alterations in hormonal responses to DA agonists in depressed patients (reviewed in Jimerson et al. 1984 b).

Since homovanillic acid (HVA) is the major CSF metabolite of brain DA, a number of workers have measured baseline levels of this metabolite in depressed patients and control subjects. In an effort to accentuate possible differences between patients and controls, some investigators incorporated patient pretreatment with probenecid, a weak acid that competitively inhibits transport of acidic amine metabolites from CSF. The accumulation of these metabolites following probenecid administration provided another measure of rate of turnover of the parent neurotransmitter (Goodwin et al. 1973).

Tables 3 and 4 list studies of CSF HVA in depressed patients with a defined control group. In an effort to sort out the varying results, the studies have been grouped according to length of drug-free interval. (Studies for which the length of the drug free interval was not specified have been included with the less-than-10-day group.) Other critical variables, such as assay methodology and measurement of probenecid levels (in probenecid studies), have been included. [Absolute values for gas chromatography/mass spectrometry (GC/MS) and fluorometric studies should not be compared directly, since the fluorometric methods apparently underestimate the total amount of CSF HVA present (Jimerson et al. 1978).]

Taking into account both baseline and postprobenecid values for HVA, five of the six studies with patients drug free for at least 10 days show decreased values for depressed patients. In the largest of these studies, the reduced HVA levels were associated with older male patients (Koslow et al. 1983). Variability in the earlier studies undoubtedly relates to the methodological problems outlined in Table 1. While most of the studies are deficient in such measures as age and sex matching of patients and controls, there is an overall trend toward reduced DA turnover in some depressed patients. Further evidence for an association between DA turnover and depressive illness emerged from data of Sedvall et al. (1980), showing that healthy volunteers with low CSF HVA [as well as those with reduced serotonin turnover (see below)] had increased incidence of affective episodes in their relatives, in comparison to volunteers with higher HVA levels. Moreover, patients with the lowest CSF HVA levels have shown the best clinical response to the DA active antidepressant nomifensine (Van Scheyen et al. 1977) and to piribedil (Post et al. 1978 b), although this observation did not hold in an antidepressant trial with bromocriptine (Nordin et al. 1981).

The studies listed in Tables 3 and 4 have reported little evidence of correlation of HVA levels with ratings of severity of depression. Suggestions of subgroup differences (e.g., bipolar vs unipolar) have not shown consistency across the CSF studies. Initially, it was proposed that retarded depressed patients had lower HVA values (Van Praag and Korf 1971; Banki 1977), which agreed with the observation that forebrain DA activity was correlated with motor activity in laboratory animals. This relationship has not been observed consistently (Papeschi and McClure 1971;

Table 3. CSF homovanillic acid (HVA) in depressed patients medication free for at least 10 days

Author	Control subjects	Assay method	Depressed[a] patients	Controls[a]	Significance of difference
Papeschi and McClure 1971	Neuro	Fluoro	19 ± 4 (17)	50 ± 6 (18)	$p < 0.001$ [b]
Goodwin et al. 1973	Nl and neuro	Fluoro	17.7± 3 (53)	22.4± 2 (28)	$p < 0.1$
			[90 ±11 (6)	194 ±24 (8)][c]	$p < 0.01$
Brodie et al. 1973	?	Fluoro	21 ± 5 (7)	51 ±11	$p < 0.05$
Berger et al. 1980	Nl	GC/MS	46.5± 6.2 (10)	44.1± 3.7 (22)	NS
				175.9±13.8 (22)][c, d	
			[sig. lower		$p < 0.01$
Banki et al. 1981	Neuro	Fluoro	28.6± 2.6 (33)	27.7± 2.2 (32)	NS
Koslow et al. 1983	Nl	GC/MS	32.4± 1.6 (49)	40.1± 2.7 (30) (male)	$p < 0.01$
			38.1± 2.1 (43)	43.7± 2.6 (32) (female)	NS

GC, gas chromatography; MS, mass spectrometry; Neuro, neurologic patients; Nl, normal volunteers

[a] HVA concentration given as nanograms per milliliter ± SEM (no. of subjects)

[b] Significance level calculated from original data using Student's t-test

[c] Values given in brackets ([]) were obtained following pretreatment of patients with probenecid

[d] Study controlled for CSF probenecid concentration

Table 4. CSF homovanillic acid (HVA) in depressed patients medication free for less than 10 days

Author	Control subjects	Assay method	Depressed[a] patients	Controls[a]	Significance of difference
Van Praag and Korf 1971	?	Fluoro	39 ± 3.6 (20)	42 ± 4.6 (12)	NS
			32 ± 2.8 (8)[e]		p<0.01
			[82 ± 9.6 (20)	91 ± 7.5 (12)[c]	NS
			[53 ±11.3 (8)][c,e]		p<0.01
Sjöström 1973	NI	Fluoro	24 ± 3 (16)	34 ± 3 (38)	p<0.1[b]
			[75 ±15% (3)	207 ±11% (14)][c,d,f]	p<0.001[b]
Van Praag et al. 1973	Neuro	Fluoro	42 ± 3.7 (28)	35 ± 3.8 (12)	NS
			[127 ±12.9 (28)	170 ±13.7 (12)][c,d]	p<0.02
Ashcroft et al. 1973	Neuro	Fluoro	20 ± 2.7 (11) UP[g]	41 ± 4.1 (31)	p<0.01
			34 ± 5.3 (9) BP[g]		NS
Takahashi et al. 1974	Neuro	Fluoro	33.9± 2.5 (30)	37.5± 3.7 (30)	NS
Bowers et al. 1974	Inmate	Fluoro	[131 ±16 (9) UP[g]	92 ± 7 (15)[c]	p<0.05[b]
Subrahmanyam 1975	NI	Fluoro	38.4± 3.4 (24)	40.2± 4 (12)	NS
Banki 1977	Neuro	Fluoro	24 ± 1.5 (55) UP[g]	33.4± 1.0 (32)	p<0.01
			15.9± 2.0 (16) BP[g]		p<0.01
Vestergaard et al. 1978	Neuro	GC	83 ± 7 (29)	45 ± 5.4 (23)	p<0.001
Ågren 1980a	NI	GC/MS	37.1±20.4 (21) UP[g]	54.7±27.3 (SD)	p<0.05
			33.3±15.8 (12) BP[g]		NS
Oreland et al. 1981	NI	GC/MS	35.7± 4.4 (6)	39.1± 3.8 (28) male	NS
				47.8± 3.5 (14) female	NS
Träskman et al. 1981	NI	GC/MS	26.1± 2.6 (8)	44.5± 3.3 (45)	p<0.01
Kasa, 1982	NI	Fluoro	22.3± 3.9 (13)	41.8± 4.2 (16)	p<0.01

GC, gas chromatography; MS, mass spectrometry
[a] HVA concentration given as nanograms per milliliter ± SEM (no. of subjects)
[b] Significance level calculated from original data using Student's t-test
[c] Values given in brackets ([ ]) were obtained following pretreatment of patients with probenecid
[d] Study controlled for CSF probenecid concentration
[e] Results calculated separately for patient subgroup with psychomotor retardation
[f] Postprobenecid HVA expressed as percentage accumulation over baseline value
[g] Mean values reported separately for unipolar (UP) and bipolar (BP) patients

Goodwin et al. 1973; Vestergaard et al. 1978). Ågren (1980a) found, however, that symptoms other than severity of depression or motor retardation may be related to DA turnover. Thus, in his study, elevated CSF HVA in unipolar depressed patients was associated with high ratings of demandingness, suspiciousness, and terminal insomnia, while low HVA was associated with initial and middle insomnia, depersonalization, and self-pity. It is unclear whether these are exclusively state-related characteristics, or underlying personality correlates of DA function. In any event, correction for such relationships might allow a closer examination of the association between the underlying depressive syndrome and CSF HVA levels.

Cerebrospinal Fluid 5-Hydroxyindoleacetic Acid in Affective Illness

There are several methodological issues particularly relevant to CSF levels of the serotonin metabolite 5-hydroxyindoleacetic acid (5-HIAA). These include aliquot chosen for 5-HIAA determination (in view of the well-described rostrocaudal gradient) and correction for body height (Åsberg et al. 1978; Bertilsson et al. 1982; Banki and Molnar 1981a). With regard to 5-HIAA, one must consider the degree to which lumbar CSF 5-HIAA reflects brain serotonin activity. Certainly, drugs thought to influence serotonergic activity alter lumbar CSF 5-HIAA in the expected direction (Post et al. 1980; Sedvall et al. 1975). However, spinal neurons may contribute 50% (Goodwin et al. 1975) or more (Aizenstein and Korf 1979) to lumbar CSF 5-HIAA. Banki and Molnar (1981b) found a high correlation ($n = 14$, $r = 0.74$, $p < 0.01$) between the 5-HIAA level in the first and last (seventh) 6-ml fraction removed during pneumoencephalography. However, Curzon et al. (1980) compared ventricular and lumbar 5-HIAA levels, finding no significant correlation.

Assay methodology must be considered. GC/MS, fluorometry, and high-pressure liquid chromatography (HPLC) with electrochemical detection have been commonly employed. The GC/MS and HPLC methods are probably more specific than the fluorometric method. These former two methods yield comparable results (M. Linnoila, personal communication).

Recent CSF 5-HIAA studies in patients with affective illness are summarized in Table 5. It is clear that no consensus exists among these authors. No author has reported a correlation of CSF 5-HIAA levels with severity of illness. In fact, several lines of indirect evidence are compatible with the hypothesis that CSF 5-HIAA levels may be state independent and genetically determined. Sedvall and Oxenstierna (1981) found a significant intraclass correlation of CSF 5-HIAA among monozygotic twins, but have not demonstrated that this concordance is higher than that found in dizygotic twins. Additionally, Swann et al. (1983) and Koslow et al. (1983) found elevated CSF 5-HIAA in 14 manic and 43 depressed women compared to the 29 female normal volunteers. Futher, Van Praag and De Haan (1980) reported that 33 of 54 depressed patients showing decreased CSF 5-HIAA accumulation after probenecid continued to demonstrate the decreased accumulation 3 months after recovery. The data of Post et al. (1980) is also compatible with 5-HIAA as a state-independent variable. However, Vestergaard et al. (1978) found that recovered patients had significantly higher ($p < 0.02$, $n = 14$, paired t-test) CSF 5-HIAA levels

Table 5. Recent CSF 5-hydroxyindoleacetic acid (5-HIAA) studies in affective illness

Author	Results	Comments
Vestergaard et al. 1978	No differences in CSF 5-HIAA for 29 depressed, 4 manic, and 33 neurological controls	Fluorometric assay. The CSF 5-HIAA in 16 recovered patients was significantly higher than their depressed values ($p < 0.02$)
Van Praag and De Haan 1980	Subgroup of patients with accumulation of 5-HIAA after probenecid	54/123 depressives had decreased 5-HIAA accumulation. Of these 33 showed this after recovery. Fluorometric assay
Berger et al. 1980	No differences before or after probenecid between 23 normals and 13 depressed patients	GC/MS method. Male subjects only
Leckman et al. 1980	No differences after probenecid between 41 patients with affective disorder, 30 with schizo-affective disorder, and 26 schizophrenics	Fluorometric assay
Ågren 1980a	Decreased 5-HIAA in 33 depressed patients	12 bipolar patients. GC/MS method
Curzon et al. 1980	5-HIAA decreased in non-agitated depressives at their worst period of depression	Decreased in 6 patients. Finding may be limited to most severe depressive period. Fluorometric method
Oreland et al. 1981	No differences in CSF 5-HIAA between 20 depressed patients and 42 controls	Suicidal patients had significantly lower CSF 5-HIAA independent of mood disorder. Recovered patients ($n = 11$) had nonsignificantly higher CSF 5-HIAA than depressives.
Korf et al. 1983	No differences between depressed and nondepressed patients; probenecid used	Nondepressed psychiatric controls used. Fluorometric assay
Swann et al. 1983	Elevated 5-HIAA in manic women compared to 29 female controls	14 patients studied. GC/MS method
Koslow et al. 1983	Elevated 5-HIAA in 43 depressed women and 5 manic women compared to 29 controls. No differences for male groups	GC/MS method. Multicenter study including 31 unipolar females and 12 bipolar females
Berrettini et al., manuscript in preparation	No differences found between 15 euthymic bipolars and 25 normal volunteers	Euthymic patients only. HPLC with electrochemical detection

GC, gas chromatography; MS, mass spectroscopy; HPLC, high-pressure liquid chromatography than depressed patients. This suggests that 5-HIAA may be state dependent. Oreland et al. (1981) found a trend in the same direction. Additional studies are required to determine whether CSF 5-HIAA levels are independent of mood state and genetically determined.

Sedvall et al. (1980) found that 28 psychiatrically healthy individuals with a positive family history of severe psychiatric illness showed greater variation in CSF 5-HIAA than did 32 normals without the positive family history. Additionally, they reported that those with abnormally low CSF 5-HIAA tended to have a family history of depression, while those with schizophrenia tended to have a higher CSF 5-HIAA. There was a significant difference between levels of CSF 5-HIAA in the relatives of schizophrenics and levels in the relatives of depressives. These data are compatible with the hypothesis that low CSF 5-HIAA is a familial state-independent vulnerability factor for affective illness. The studies Van Praag and De Haan (1980) are also compatible with this hypothesis. They found that some depressed individuals with low postprobenecid accumulation of CSF 5-HIAA still showed this low accumulation after recovery from the episode of depression. These individuals (Van Praag 1980) tended to respond well to oral 5-hydroxytryptophan treatment (only one of 13 relapsed) compared to those individuals with normal postprobenecid 5-HIAA accumulation (five of seven relapsed). Additionally, Van Praag and De Haan (1980) found a greater number of hospital admissions for depression among the low 5-HIAA groups and among their relatives compared to the normal 5-HIAA group. However, most importantly there was no report of higher incidence of illness among the relatives of the low 5-HIAA group. These reports require confirmation by other investigators before they can be accepted as providing evidence for a serotonergic vulnerability factor in affective illness.

The report by Van Praag (1980) that low 5-HIAA depressives respond well to "serotonergic-potentiating" compounds is similar to results of Åberg-Wistedt et al. (1982) showing that low CSF 5-HIAA (and HVA) predicted patient responsiveness to the serotonin-selective antidepressant zimelidine. Dahl et al. (1982) found that both femoxitine and desipramine lowered CSF 5-HIAA in 17 depressed patients. Pretreatment 5-HIAA levels did not predict response to either drug in this double-blind study.

One aspect of CSF 5-HIAA studies in affective illness is the relationship of low 5-HIAA to suicide attempts. Åsberg et al. (1976) first reported low CSF 5-HIAA to be associated with violent suicide attempts. This observation has been supported by Ågren (1980b), Brown et al. (1979, 1982), Oreland et al. (1981), and Träskman et al. (1981). These studies are fascinating because they delineate a correlation between a biochemical variable and human behavior independent of syndromal psychiatric diagnosis. The studies by Brown et al. (1979, 1982) and by Oreland et al. (1981) included subjects who were suicide attempters without prominent depressive symptoms. Most carried diagnoses of personality disorder. These studies again suggest that certain biochemical characteristics may predispose one toward a defined complex behavior. More generally, these reports may indicate that one should study groups of psychiatric patients with certain behavior patterns rather than groups of patients who are diagnostically homogeneous.

In our study of euthymic bipolar patients, mean 5-HIAA levels for six patients with history of suicide attempt did not differ from the rest of the patient group or the control group (Berrettini et al., work in preparation). However, this is a small strictly bipolar sample in which there was only one violent attempter. Roy-Byrne et al. (1983) in their study of somewhat atypical bipolar and unipolar patients did not find lower CSF 5-HIAA among the suicide attempters ($n=24$). Three-fourths of their patients were bipolar. Additionally, Ågren (1983) has reported that CSF 5-HIAA is unrelated to suicide in bipolar patients. Thus bipolars may be representative of a separate group of affective disorders as far as 5-HIAA and suicide are concerned. From the genetic perspective, however, unipolar and bipolar patients seem to share a common diathesis (Gershon et al. 1982). Low 5-HIAA has also been noted in schizophrenic suicide attempters (Ninan et al. 1983). Vestergaard et al. (1978) did not find lower CSF 5-HIAA to be associated with suicide attempts in a study of both bipolar and unipolar patients. These mixed results suggest that the association between reduced 5-HIAA and a history of suicide attempts may be obscured by other state- and diagnosis-related influences on CSF 5-HIAA.

In conclusion, recent studies of CSF 5-HIAA in untreated affective disorders have not provided consistent results. There are reports which are compatible with the hypothesis that some depressions are associated with a state-independent familial serotonergic deficit, but these results are tentative pending replication. The suicide studies are more consistent. They suggest that unipolar and personality disorder suicide attempters have low CSF 5-HIAA. This finding is very interesting because it correlates a biochemical variable with a highly complex specific human behavior.

Overview

In summary, the CSF amine metabolite studies reviewed above have not singled out replicable evidence for a specific alteration of catecholamine or serotonin turnover in major depressive illness. Increased awareness of potentially confounding methodological variables may improve consistency in future studies. Discovery of an isolated alteration in the turnover of a single amine neurotransmitter in depressive illness seems, however, relatively unlikely. Association between neurotransmitter function and state and trait personality variables opens exciting new directions for future CSF metabolite studies. The association between reduced CSF 5-HIAA and suicidal behavior represents a relatively robust thrust in this direction. Given the growing understanding of functional regulation of brain neurotransmitter systems, future CSF metabolite studies will benefit from evaluation of the interplay of multiple transmitters, receptor sensitivity, neuromodulators, and postreceptor effector systems.

References

Åberg-Wistedt A, Ross SB, Jostell K-G, Sjöquist B (1982) A double-blind study of zimelidine, a serotonin uptake inhibitor, and desipramine, a noradrenaline uptake inhibitor, in endogenous depression: II. Biochemical findings. Acta Psychiatr Scand 66:66

Ågren H (1980a) Symptom patterns in unipolar and bipolar depression correlating with monoamine metabolites in the cerebrospinal fluid: I. General patterns. Psychiatry Res 3:211

Ågren H (1980b) Symptom patterns in unipolar and bipolar depression correlating with monoamine metabolites in the cerebrospinal fluid: II. Suicide. Psychiatry Res 3:225

Ågren H (1983) Life at risk: markers of suicidality in depression. Psychiatr Dev 1:87

Aizenstein ML, Korf J (1979) On the elimination of centrally formed 5-HIAA by CSF and urine. J Neurochem 32:1227

Antelman SM, Chiodo LA (1981) Dopamine autoreceptor subsensitivity: a mechanism common to the treatment of depression and the induction of amphetamine psychosis. Biol Psychiatry 16:717

Åsberg M, Träskman L, Thorén P (1976) 5-HIAA in the CSF: a biochemical suicide predictor? Arch Gen Psychiatry 33:1196

Åsberg M, Bertilsson L, Thorén P, Traskman L (1978) CSF monoamine metabolites in depressive illness. In: Garattini S (ed) Depressive disorders. Schattauer, Stuttgart

Ashcroft GW, Blackburn IM, Eccleston D, Glen AIM, Hartley W, Kinloch NE, Lonergan M, Murray LG, Pullar IA (1973) Changes on recovery in the concentrations of tryptophan and the biogenic amine metabolites in the cerebrospinal fluid of patients with affective illness. Psychol Med 3:319

Ashcroft GW, Dow RC, Yates CM, Pullar IA (1975) Significance of lumbar CSF metabolite measurements in affective illness. In: Tuomisto J, Paasonen MK (eds) Proceedings of the 6th International congress of pharmacology vol 3. Finnish Pharmacological Society, Helsinki, p 277

Ballenger JC, Post RM, Jimerson DC, Lake CR, Murphy D, Zuckerman M, Cronin C (1984) Biochemical correlates of personality traits in normals: an exploration study. Personality and individual differences (in press)

Banki CM (1977) Correlation between cerebrospinal fluid amine metabolites and psychomotor activity in affective disorders. J Neurochem 28:255

Banki CM, Molnar G (1981a) The influence of age, height and body weight on CSF amine metabolites and tryptophan in women. Biol Psychiatry 16:753

Banki CM, Molnar G (1981b) CSF 5-HIAA as an index of central serotonergic processes. Psychiatry Res 5:23

Banki CM, Molnar G, Fekete I (1981) Correlation of individual symptoms and other clinical variables with cerebrospinal fluid amine metabolites and tryptophan in depression. Arch Psychiatr Nervenkr 229:345

Berger PA, Faull KF, Kilkowski J, Anderson PJ, Kraemer H, Davis KL, Barchas JD (1980) CSF monoamine metabolites in depression and schizophrenia. Am J Psychiatry 137:174

Bertilsson L, Åsberg M, Lantto O, Scalia-Tomba G-P, Traskman-Bendz L, Tybring G (1982) Gradients of monoamine metabolites and cortisol in CSF of psychiatric patients and healthy controls. Psychiatry Res 6:77

Bowers MB (1974) Lumbar CSF 5-hydroxyindoleacetic acid and homovanillic acid in affective syndromes. J Nerv Ment Dis 158:325

Brodie HKH, Sack R, Siever L (1973) Clinical studies of L-5-hydroxytryptophan in depression. In: Barchas J, Usdin E (eds) Serotonin and behavior. Academic, New York, p 549

Brown GL, Goodwin FK, Ballenger JC, Goyer PF, Major LF (1979) Aggression in humans correlates with CSF amine metabolites. Psychiatry Res 1:131

Brown GL, Ebert MH, Goyer PF, Jimerson DC, Klein WJ, Bunney WE, Goodwin FK (1982) Aggression, suicide and serotonin: relationships to CSF amine metabolites. Am J Psychiatry 139:741

Bunney WE Jr, Davis JM (1965) Norepinephrine in depressive reactions. Arch Gen Psychiatry 13:483

Charney DS, Menkes DB, Heninger GR (1981) Receptor sensitivity and the mechanism of action of antidepressant treatment: implications for the etiology and therapy of depression. Arch Gen Psychiatry 38:1160

Charney DS, Heninger GR, Sternberg DE, Landis H (1982) Abrupt discontinuation of tricyclic antidepressant drugs: evidence for noradrenergic hyperactivity. Br J Psychiatry 141:377

Chase TN, Gordon EK, Ng LKY (1973) Norepinephrine metabolism in the central nervous system of man: studies using 3-methoxy-4-hydroxyphenylethylene glycol levels in cerebrospinal fluid. J Neurochem 21:581

Christensen NJ, Vestergaard P, Sorensen T, Rafaelsen OJ (1980) Cerebrospinal fluid adrenaline and noradrenaline in depressed patients. Acta Psychiatr Scand 61:178

Colonna L, Petit M, Lepine JP (1979) Bromocriptine in affective disorders: a pilot study. J Affective Disord 1:173

Curzon G, Kantamaneni BD, Van Boxel P, Gillman PK, Bartlett JR, Bridges PK (1980) Substances related to 5-hydroxytryptamine in plasma and in lumbar and ventricular fluids of psychiatric patients. Acta Psychiatr Scand [Suppl] 280:3

Dahl L-E, Lundin L, Le Fèvre Honoré, Dencker SJ (1982) Antidepressant effect of femoxetine and desipramine and relationship to the concentration of amine metabolites in cerebrospinal fluid: a double-blind evaluation. Acta Psychiatr Scand 66:9

Elsworth JD, Redmond DE Jr, Roth RH (1982) Plasma and cerebrospinal fluid 3-methoxy-4-hydroxyphenylethylene glycol (MHPG) as indices of brain norepinephrine metabolism in primates. Brain Res 235:115

Garver DL, Davis JM (1979) Biogenic amine hypotheses of affective disorders. Life Sci 24:383

Gershon ES, Hamovit J, Guroff JJ, Dibble E, Leckman JF, Sceery W, Targum SD, Nurnberger JJ, Goldin LR, Bunney WE Jr (1982) A family study of schizoaffective, bipolar I, bipolar II, unipolar and normal control probands. Arch Gen Psychiatry 39:1157

Goodwin FK, Post RM, Dunner DL, Gordon EK (1973) Cerebrospinal fluid amine metabolites in affective illness: the probenecid technique. Am J Psychiatry 130:73

Goodwin FK, Post RM, Jimerson DC (1975) Studies of CSF amine metabolites in affective illness and schizophrenia: In: Tuomisto J, Paasonen MK (eds) Proceedings of the 6th international congress of pharmacology, vol 3. Finnish Pharmacological Society, Helsinki, p 285

Halaris A, Belendiuk K, Freedman DX (1975) Antidepressant drugs affect dopamine uptake. Biochem Pharmacol 24:1896

Jimerson DC, Post RM, Goodwin FK (1976) Antidepressant treatments and alterations in central serotonin turnover in affective illness. Monogr Neural Sci 3:15–22

Jimerson DC, Gordon EK, Post RM, Goodwin FK (1978) Homovanillic acid in human CSF: Comparison of fluorimetry and gas chromatography – mass spectrometry. Communi Psychopharmacol 2:343

Jimerson DC, Ballenger JC, Lake CR, Post RM, Goodwin FK, Kopin IJ (1981) Plasma and CSF MHPG in normals. Psychopharmacol Bull 17:86

Jimerson DC, Rubinow DR, Ballenger JC, Post RM, Kopin IJ (1984a) CSF norepinephrine metabolites in depressed patients: new methodologies. In: Usdin E, Carlsson A, Dahlstrom A and Engel J (eds) Catecholamines, Part C: Neuropharmacology and Central Nervous System – Therapeutic Aspects. Alan R. Liss, New York, p 123

Jimerson DC, Cutler NR, Post RM, Rey A, Gold PW, Brown GM, Bunney WE Jr (1984b) Neuroendocrine responses to apomorphine in depressed patients and healthy controls. Psychiatr Res (in press)

Kasa K, Otsuki S, Yamamoto M, Sato M, Kuroda H, Ogawa N (1982) Cerebrospinal fluid γ-aminobutyric acid and homovanillic acid in depressive disorders. Biol Psychiatry 17:877

Kessler JA, Fenstermacher JD, Patlak CS (1976) 3-Methoxy-4-hydroxyphenylethyleneglycol (MHPG) transport from the spinal cord during spinal subarachnoid perfusion. Brain Res 102:131

Kopin IJ, Gordon EK, Jimerson DC, Polinsky RJ (1983) Relation between plasma and cerebrospinal fluid levels of 3-methoxy-4-hydroxyphenylglycol. Science 219:73

Korf J, van den Burg W, van den Hoofdakker RH (1983) Acid metabolites and precursor amino acids of 5-hydroxytryptamine and dopamine in affective and other psychiatric disorders. Psychiatr Clin 16:1

Koslow SH, Maas JW, Bowden CL, Davis JM, Hanin I, Javaid J (1983) CSF and urinary biogenic amines and metabolites in depression and mania: a controlled, univariate analysis. Arch Gen Psychiatry 40:999

Lake CR, Pickar D, Ziegler MG, Lipper S, Slater S, Murphy DL (1982) High plasma norepinephrine levels in patients with major affective disorder. Am J Psychiatry 139:1315

Leckman JF, Cohen DJ, Shayvitz BA, Caparulo BK, Heninger GR, Bowers MB Jr (1980) CSF monoamine metabolites in child and adult psychiatric patients. Arch Gen Psychiatry 37:677

Lee T, Tang SW (1982) Reduced presynaptic dopamine receptor density after chronic antidepressant treatment in rats. Psychiatry Res 7:111

Meek JL, Neff NJ (1973) The rate of formation of 3-methoxy-4-hydroxyphenethyleneglycol sulfate in brain as an estimate of the rate of formation of norepinephrine. J Pharmacol Exp Ther 184:570

Murphy DL, Brodie HKH, Goodwin FK, Bunney WE Jr (1971) L-dopa: regular induction of hypomania in "bipolar" manic depressive patients. Nature 229:135

Ninan PT, Van Kammen DP, Scheinen M, Linnoila M, Bunney WE, Goodwin FK (1984) Cerebrospinal fluid 5-HIAA in suicidal schizophrenic patients. Am J Psychiatry 141:566

Nordin C, Siwers B, Bertilsson L (1981) Bromocriptine treatment of depressive disorders: clinical and biochemical effects. Acta Psychiatr Scand 64:25

Oreland L, Wiberg Å, Åsberg M, Träskman L, Sjöstrand L, Thorén P, Bertilsson L, Tybring G (1981) Platelet MAO activity and monoamine metabolites in cerebrospinal fluid in depressed and suicidal patients and in healthy controls. Psychiatry Res 4:21

Papeschi R, McClure DJ (1971) Homovanillic and 5-hydroxyindoleacetic acid in cerebrospinal fluid of depressed patients. Arch Gen Psychiatry 25:354

Polinsky RJ, Jimerson DC, Kopin IJ (1984) MHPG levels in plasma and CSF in patients with chronic autonomic failure. Neurology (in press)

Post RM, Gordon EK, Goodwin FK, Bunney WE (1973) Central norepinephrine metabolism in affective illness: MHPG in the cerebrospinal fluid. Science 179:1002

Post RM, Lake CR, Jimerson DC, Bunney WE Jr, Wood JH, Ziegler MG, Goodwin FK (1978a) Cerebrospinal fluid norepinephrine in affective illness. Am J Psychiatry 135:907

Post RM, Gerner RH, Carmen JS, Gillin JC, Jimerson DC, Goodwin FK, Bunney WE Jr (1978b) Effects of a dopamine agonist piribedil in depressed patients: relationship of pretreatment homovanillic acid to antidepressant response. Arch Gen Psychiatry 35:609

Post RM, Ballenger JC, Goodwin FK (1980) Cerebrospinal fluid studies of neurotransmitter function in manic and depressive illness. In: Wood JH (ed) Neurobiology of cerebrospinal fluid vol 1. Plenum, New York, p 685

Post RM, Ballenger JC, Jimerson DC, Bunney WE Jr (to be published) Plasma MHPG is inversely correlated with depression, hypochondriasis, and psychasthenia scores on the MMPI in normal subjects

Randrup A, Braestrup C (1977) Uptake inhibition of biogenic amines by newer antidepressant drugs: relevance to the dopamine hypothesis of depression. Psychopharmacology (Berlin) 53:309

Randrup A, Munkvad I, Fog R, Gerlach J, Molander L, Kjellberg B, Scheel-Krüger J (1975) Mania, depression, and brain dopamine. Curr Dev Psychopharmacol Spectrum 2:205

Roy-Byrne P, Post RM, Rubinow DR, Linnoila M, Savard R, Davis D (1983) CSF 5-HIAA and personal and family history of suicide in affectively ill patients: a negative study. Psychiatry Res 10:263

Schildkraut JJ (1965) The catecholamine hypothesis of affective disorders: a review of supporting evidence. Am J Psychiatry 122:509

Schildkraut JJ (1978) Current status of the catecholamine hypothesis of affective disorders. In: Lipton MA, DiMascio A, Killam KF (eds) Psychopharmacology: a generation of progress. Raven, New York, p 1223

Sedvall G, Oxenstierna G (1981) Genetic and environmental influences on central monoaminergic mechanisms in man. Presented at the 3rd World congress of biological psychiatry, Stockholm (Abstract No 5344)

Sedvall G, Alfredsson G, Bjerkenstedt L, Eneroth P, Fyrö B, Härnryd C, Swahn GG (1975) Selective effects of psychoactive drugs on levels of monoamine metabolites and prolactin in cerebrospinal fluid of psychiatric patients. In: Tuomisto J, Paasonen MK (eds) Proceedings of the 6th international congress of pharmacology. Finnish Pharmacological Society, Helsinki, p 255

Sedvall G, Fyrö B, Gullberg B, Nybäck H, Wiesel F-A, Wode-Helgodt B (1980) Relationships in healthy volunteers between concentrations of monoamine metabolites in cerebrospinal fluid and family history of psychiatric morbidity. Br J Psychiatry 136:366

Serra G, Argiolas A, Klimek V, Fadda F, Gessa GL (1979) Chronic treatment with antidepressants prevents the inhibitory effect of small doses of apomorphine on dopamine synthesis and motor activity. Life Sci 25:415

Shaw DM, O'Keeffe R, MacSweeney DA, Brooksbank BWL, Noguera R, Coppen A (1973) 3-Methoxy-4-hydroxyphenylglycol in depression. Psychol Med 3:333

Shopsin B, Gershon S (1978) Dopamine receptor stimulation in the treatment of depression: piribedil (ET-495). Neuropsychobiology 4:1

Shopsin B, Wilk S, Sathananthan G, Gershon S, Davis K (1974) Catecholamines and affective disorders revised: a critical assessment. J Nerv Ment Disord 158:369

Sjöström R (1973) Cerebrospinal fluid content of 5-hydroxyindoleacetic acid and homcvanillic acid in manic-depressive psychosis. Acta Univ Upsaliensis 154:5

Subrahmanyam S (1975) Role of biogenic amines in certain pathological conditions. Brain Res 87:355

Swann AC, Secunda S, Davis JM, Robins E, Hanin I, Koslow SH, Maas JW (1983) CSF monoamine metabolites in mania. Am J Psychiatry 140:396

Takahashi S, Yamane H, Kondo H, Tani N, Kato N (1974) CSF monoamine metabolites in alcoholism: a comparative study with depression. Folia Psychiatr Neurol Jpn 28:347

Träskman L, Åsberg M, Bertilsson L, Sjöstrand L (1981) Monoamine metabolites in CSF and suicidal behavior. Arch Gen Psychiatry 38:631

Van Praag HM (1977) Significance of biochemical parameters in the diagnosis, treatment and prevention of depressive disorders. Biol Psychiatry 12:101

Van Praag HM (1980) Central monoamine metabolism in depressions. I. Serotonin and related compounds. Compr Psychiatry 21:30

Van Praag HM, De Haan S (1980) Central serotonin deficiency – a factor which increases depression vulnerability? Acta Psychiatr Scand [Suppl] 280:89

Van Praag HM, Korf J (1971) Retarded depression and the dopamine metabolism. Psychopharmacologia 19:199

Van Praag HM, Korf J, Schut D (1973) Cerebral monoamines and depression: an investigation with the probenecid technique. Arch Gen Psychiatry 28:827

Van Scheyen JD, Van Praag HM, Korf J (1977) Controlled study comparing nomifensine and clomipramine in unipolar depression, using the probenecid technique. Br J Clin Pharmacol 4:179S

Vestergaard P, Sorensen T, Hoppe E, Raphaelsen OJ, Yates CM, Nicolaou N (1978) Biogenic amine metabolites in cerebrospinal fluid of patients with affective disorders. Acta Psychiatr Scand 58:88

Waehrens J, Gerlach J (1981) Bromocriptine and imipramine in endogenous depression: a double-blind controlled trial in out-patients. J Affective Disord 3:193

Wilk S, Shopsin B, Gershon S, Suhl M (1972) Cerebrospinal fluid levels of MHPG in affective disorders. Nature 235:440

CSF Studies in Schizophrenia: A Multidimensional Approach

Wagner F. Gattaz, T. Gasser, and H. Beckmann

Introduction

The aim of the present study was the simultaneous investigation of different neuronal systems in a group of schizophrenic patients and healthy controls in order to investigate their possible role in the disease, their interrelationship, and the effects of neuroleptic drugs upon their function. We have investigated the brain metabolism through the determination of the concentrations of hormones, neurotransmitters, and their major metabolites in the cerebrospinal fluid (CSF).

Partial results from this investigation have been published separately elsewhere (Beckmann et al. 1982; Beckmann et al., to be published; Gattaz et al. 1982a, b, 1983a–c). They are summarized in Table 1 and the results of simultaneous statistical evaluation through multidimensional scaling (MDS) are in Fig. 1.

Patients and Methods

The study involved 28 paranoid schizophrenic patients (all males mean age 30.6 ± 8.0 years) and 16 controls (14 males and two females, mean age 35.0 ± 15.7 years).

Patients were diagnosed according to the Research Diagnostic Criteria (Spitzer et al. 1975). Two experienced psychiatrists independently evaluated the patients' psychopathological state by means of the Brief Psychiatric Rating Scale (BPRS). Fifteen patients were under treatment with neuroleptic drugs (butyrophenones and phenothiazines) for at least 3 weeks (mean dosis $\pm$ SD in chlorpromazine equivalents $= 585 \pm 755$ mg/day). Thirteen patients did not take any drug for a period of at least 4 weeks prior to the study. Both subgroups of patients were on a free standardized diet.

Controls were subjects with nonspecific neurological symptomatology (headaches, dizziness, etc.) in whom a lumbar puncture was necessary for diagnostic purposes. Controls were not under drug treatment at the time of the lumber puncture.

Cerebrospinal fluid was obtained by lumbar puncture with probands in a sitting position between 9 and 10 a.m., after they had fasted for 12 h and had had bed rest for 10 h. Sixteen ml CSF was removed without additions. To avoid rostral-caudal gradient effects, samples were gently mixed and then immediately frozen on dry ice and stored in a freezer at $-70 \,^\circ$C until analyzed.

The biochemical determinations were performed blind to the origin of each sample. Nonparametric tests were used for the statistical evaluation of the data.

The Serotonergic System

The major metabolite of serotonin (5-HT) in the CSF is 5-hydroxyindoleacetic acid (5-HIAA). The concentrations of 5-HIAA in the CSF have been shown to reflect to a considerable extent the metabolism of the parent amine 5-HT in the brain (Asberg et al. 1976; Banki and Molnár 1980; Sedvall et al. 1975).

Studies in schizophrenic patients have been contradictory so far. Sedvall and Wode-Helgodt (1980) reported that high or deviant concentrations of 5-HIAA were significantly related to family history of schizophrenia, whereas Winblad et al. (1979) found lower mean values of 5-HT and 5-HIAA in different brain areas of deceased schizophrenic patients.

In our studies schizophrenic patients showed lower concentrations of 5-HIAA than controls ($p < 0.005$). No differences were found in the levels of the metabolite between patients with and without neuroleptics. 5-HIAA correlated positively with the items "grandiosity" ($p < 0.05$) and "hallucinatory behavior" ($p < 0.01$) from the score "thought disturbances."

Our finding of reduced 5-HIAA in schizophrenic patients is in agreement with some investigations (Ashcroft et al. 1966; Bowers et al. 1969) but not with others (Bowers 1973; Gomes et al. 1980; Persson and Roos 1969; Post et al. 1973). It seems unlikely that this reduction is an artifact due to a drug effect, as no significant difference was found in the 5-HIAA concentrations between patients with and without neuroleptics, which is in line with six out of seven studies in the literature (see review: Gattaz et al. 1982 b).

The possibility that our results reflect a decrease in synthesis of 5-HT is supported by the findings of Winblad et al. (1979) of diminished levels of this amine in brain tissues of schizophrenics. In this regard, it is of interest to note that drugs capable of producing psychotic states in vulnerable individuals, such as monoamine oxidase inhibitors, tricyclic antidepressants, and LSD, all decrease the firing rates of 5-HT-containing neurons.

This, together with the findings in depressed and suicidal patients, indicates that a disordered serotonergic function is more likely to be nonspecifically related to some psychopathological states or manifestations rather than to specific nosological entities.

The Noradrenergic System

A role for the noradrenergic system in schizophrenia has been suggested (Stein and Wise 1971). Noradrenaline (NA) has recently been reported to be increased in the CSF of schizophrenic patients (Gomes et al. 1980; Lake et al. 1980; Sternberg et al. 1981), which is in line with earlier findings of increased NA in the brains of deceased schizophrenics (Bird et al. 1979; Farley et al. 1978). The view has been put forward that an increased noradrenergic activity could underline vulnerability to the disease.

To investigate this question further we determined the CSF concentrations of NA and 3-methoxy-4-hydroxyphenylglycol (MHPG), its major metabolite in the brain. Furthermore, we investigated the CSF concentrations of cortisol, release of which is supposed to be under noradrenergic tonic suppression (Kizer and Youngblood 1978).

Noradrenaline and 3-Methoxy-4-hydroxyphenylglycol

No significant difference in NA concentration were found between patients not receiving neuroleptics and controls. Conversely, patients under neuroleptic therapy showed significantly higher CSF concentrations of NA than controls ($p < 0.01$).

No significant differences could be detected in the CSF concentrations of MHPG among patients receiving and those not receiving neuroleptics and controls.

The finding of normal NA and MHPG concentrations in schizophrenic patients not receiving neuroleptics does not support the hypothesis of a disturbed noradrenergic function in schizophrenia. Our results indicate further that neuroleptic treatment might enhance the concentrations of NA in the CSF. This agrees with animal studies where neuroleptics were shown to increase the concentration of NA in different areas of the brain, probably through a positive feedback mechanism after blockade of pre- or postsynaptic receptors (Bartholini et al. 1976). These data taken together suggest that the findings of increased NA in the CSF and brain areas of deceased schizophrenic patients, as described in the literature, could be due at least in part to the chronic intake of neuroleptic drugs.

Cortisol

No significant differences in cortisol concentration were found between patients not receiving neuroleptics and controls. Patients under drug therapy showed significantly higher cortisol concentration than patients not receiving neuroleptics ($p < 0.05$). The cortisol concentrations correlated positively with the item "anxiety" from the BPRS ($p < 0.05$).

As expected, the blockade of noradrenergic receptors by neuroleptic drugs lead to an increased cortisol release in the circulation and consequently in the CSF, as this hormone readily crosses the blood-brain barrier. The concentrations of cortisol in the CSF therefore correlate very highly with the concentrations of plasma free cortisol (Murphy et al. 1967).

High concentrations of cortisol have repeatedly been correlated with depressive disorders (Carroll 1976). On the other hand, the occurrence of depressive states in schizophrenics after the remission of the psychosis has been described by several authors (Heinrich 1976; Helmchen and Hippius 1967). Although other factors cannot be excluded, at least one part of these postpsychotic depressive states is assumed to be a consequence of neuroleptic treatment (pharmacogenic depression)

(McGlashan and Carpenter 1976). In view of these data, our finding of increased cortisol after neuroleptics could provide a biological substrate for some postpsychotic depressive states in schizophrenics.

In this regard, the correlation between the cortisol concentrations and the item "anxiety" in our study is of interest, as the latter may be seen as a frequent manifestation in depressed states.

The Dopaminergic System

The dopamine (DA) hypothesis is presently the dominant theory concerning the etiology of schizophrenia (Snyder 1976; Snyder et al. 1974). A hypothetical hyperactivity of the dopaminergic system in schizophrenia has been proposed, on the basis of the pharmacologic evidence that:

1. Virtually all antipsychotic drugs have a dopamine-receptor-blocking effect
2. DA agonists, e.g., amphetamine, are likely to produce a schizophrenia-like psychosis in certain individuals.

In order to investigate the dopaminergic activity in our sample we determined the CSF concentrations of DA and its major metabolite homovanillic acid (HVA), as well as the concentrations of prolactin, release of which from the anterior pituitary gland is tonically inhibited by the tuberoinfundibular DA neurons.

Dopamine and Homovanillic Acid

No significant differences in DA concentration were found between patients not receiving neuroleptics and controls. Patients on neuroleptics had significantly higher DA concentrations than controls ($p < 0.001$) and patients not receiving neuroleptics ($p < 0.01$). No significant differences in HVA concentration were found among the three subgroups.

These results do not support the hypothesis of an increased DA turnover in schizophrenic patients. On the other hand, the finding of increased DA concentrations in patients on neuroleptics is in line with animal experiments (Bartholini et al. 1976; Stadler et al. 1975) and could provide a rationale for at least part of the reports on increased DA concentrations in the brain of deceased schizophrenic patients (Bird et al. 1977, 1979).

Prolactin

No significant differences in the concentration of prolactin were found between patients without neuroleptics and controls. As expected, patients on neuroleptics showed higher prolactin levels than controls ($p < 0.001$) and than patients without neuroleptics ($p < 0.001$).

The finding of normal concentrations of prolactin in schizophrenics without drugs indicates that a hyperdopaminergic state in schizophrenia, so far as it exists, does not affect the tuberoinfundibular dopaminergic pathway which controls the secretion of prolactin from the anterior pituitary.

Summarizing, our results from the determinations of DA, HVA, and prolactin do not support the hypothesis of an increased DA turnover in schizophrenia. The possibility that a hyperdopaminergic function in this disease is caused by an increase in the number of DA receptors or an enhanced DA receptor sensibility has been investigated by other autors (Crow et al. 1980; Langer et al. 1981) and up to now no conclusive support for the DA hypothesis in this simplest form has been obtained.

The Cholinergic System

A functional relationship between the cholinergic and dopaminergic system has been suggested (Barbeau 1962). This was mainly based on the fact that anticholinergic drugs might improve some symptoms of Parnkinson's disease, a condition characterized by a DA deficiency in the striatum.

The short half-life of acetylcholine (ACh) in body fluids makes it difficult to measure the concentrations of this transmitter in the CSF. A way out of this difficulty is to determine the concentrations of cyclic guanosine 3'5'-monophosphate (cGMP), which has been shown to reflect central cholinergic activity (Ebstein et al. 1976; Smith et al. 1976).

The CSF concentrations of cGMP were significantly reduced in the subgroup of patients not receiving neuroleptics as compared to controls ($p < 0.01$). In the group of patients under neuroleptic therapy this difference was just below the level of significance ($p = 0.04$). There was a positive correlation between cGMP levels and the number of psychiatric hospital admissions; the latter thus indicates indirectly the amount of drug intake during the course of the illness.

The finding of reduced cGMP in schizophrenics could suggest a reduced cholinergic activity in these patients. Several studies pointed to a functional balance between the cholinergic and the dopaminergic systems (Andén and Bedard 1971; Bartholini et al. 1973). Thus, even in the face of a normal dopaminergic activity in schizophrenia, a reduced cholinergic activity would result in a *relative* hyperfunction of the dopaminergic system. We speculate whether this cholinergic-dopaminergic imbalance could be viewed as a possible biological factor underlying vulnerability to schizophrenia. In this context, the tendency shown by neuroleptics to increase cGMP concentrations observed in our and other (Ebstein et al. 1976) studies could be understood as a restoration of the cholinergic-dopaminergic balance through cholinergic stimulation and dopaminergic blockade. This assumption is in line with laboratory experiments which showed an increased ACh release in the brain and CSF after the administration of neuroleptics (Stadler et al. 1973; Trabucchi et al. 1975). Interestingly, amphetamines, which can produce a schizophrenia-like psychosis, have been found to work in the opposite direction, markedly reducing the ACh turnover in the brain (Trabucchi et al. 1975).

The Glutamatergic System

Glutamate is a dicarboxylic amino acid with the role of an excitatory neurotransmitter in the CNS. Moreover, it functions as a detoxifying substrate, trapping ammonia to form glutamine, and further as the precursor of gamma-aminobutyric acid. The release of glutamate can be inhibited by dopaminergic neurons, whereas the glutamatergic system itself activates the cholinergic activity (Johnson 1972).

Patients not receiving neuroleptics and controls showed similar concentrations of glutamate in the CSF. Conversely, patients on neuroleptics showed significantly higher CSF concentrations of glutamate than controls ($p < 0.001$) and patients not receiving neuroleptics ($p < 0.01$). The concentration of glutamate correlated positively with the number of psychiatric hospital admissions ($p < 0.05$).

Our findings indicate that the blockade of dopamine receptors by neuroleptic treatment may enhance the concentrations of glutamate, which could be expected from animal studies (Rowlands and Roberts 1980). Taking into account the excitatory effect of glutamatergic neurons upon the cholinergic system, we speculate whether the enhanced glutamate concentrations after neuroleptics could be of importance for the restoration of the cholinergic-dopaminergic balance in schizophrenia.

The Simultaneous Evaluation of the Data Through Multidimensional Scaling

We reported above our findings of differences in the concentrations of several substances related to central neuronal activity among schizophrenics receiving and those not receiving neuroleptics and controls. Although statistically significant, none of these differences was strong enough to separate the three subgroups without a considerable overlap.

We speculate whether these isolated findings are all the manifestation of the same basic biological process. In this case, the simultaneous evaluation of all variables together should provide more information than each of them taken alone. This hypothesis could be verified by obtaining a two-dimensional representation via MDS of the subjects represented in a high-dimensional space of 17 biochemical parameters.

Statistical methods have been described in detail elsewhere (Gasser and Möcks 1983). Briefly, in order to make the 17 biochemical parameters comparable, they were standardized for all subjects such that the standard deviation with respect to the control group became 1. The full data set can be regarded as a cloud of points consisting of the 44 subjects in the 17-dimensional parameter space. MDS seeks a two-dimensional representation of this 17-dimensional cloud of points, while retaining as far as possible the distances between subjects. No information about the composition of the groups is used. The MDS algorithm is based on the ranking of the distances and this lowers the influence of an aberrant value in some dimension. To characterize the normative region occupied by the control group, $(1 - p)$ convex

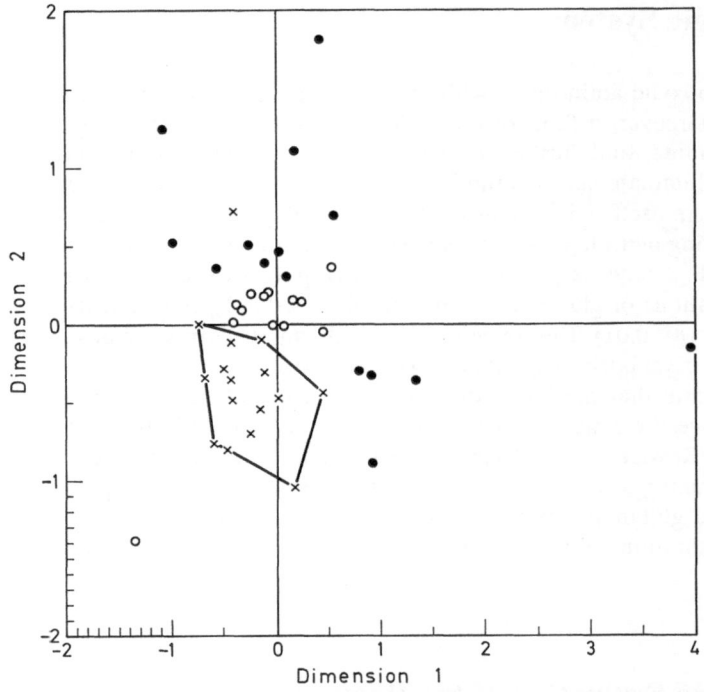

Fig. 1. MDS plot of 17 CSF parameters in 13 drug-free schizophrenics (○), 15 schizophrenics on drugs (●), and 16 controls (×)

Table 1. Concentrations of 17 substances in the cerebrospinal fluid of 28 schizophrenic patients (13 off drugs and 15 on drugs) and 16 controls. Mean ± SD (Median)

	Controls	Patients on drugs	Patients of drugs	
Glutamate	18.9± 5.6 (19.4)	25.2± 5.8 (26.6)[a]	20.5± 4.0 (19.4)	(nmol/ml)
Dopamine	207.0±263.0 (119.0)	524.0±549.0 (286.0)[a]	200.0±124.0 (151.0)	(pg/ml)
Noradrenaline	133.0± 37.0 (124.0)	210.0± 73.0 (224.0)[a]	163.0± 44.0 (158.0)	(pg/ml)
Adrenaline	159.0± 85.0 (85.0)	325.0±246.0 (330.0)[a]	225.0±118.0 (230.0)	(pg/ml)
HVA	33.9± 18.1 (18.1)	39.0± 15.1 (39.6)	37.3± 20.2 (32.9)	(ng/ml)
MHPG	16.2± 4.9 (16.7)	14.6± 3.4 (14.4)	17.8± 4.8 (18.7)	(ng/ml)
HIAA	15.3± 4.4 (16.3)	9.9± 3.7 (8.6)[a]	11.8± 4.6 (11.3)[a]	(ng/ml)
PAA (free)	22.2± 11.8 (21.2)	14.5± 10.4 (12.8)	11.5± 5.7 (10.8)[a]	(ng/ml)
PAA (conj.)	23.6± 20.8 (23.6)	14.8± 7.8 (13.8)	20.4± 19.8 (18.2)	(ng/ml)
Cyclic AMP	15.2± 4.2 (14.5)	11.3± 4.3 (9.2)[a]	10.4± 9.4 (7.5)[a]	(pmol/ml)
Cyclic GMP	4.0± 1.2 (4.2)	2.8± 1.2 (2.6)[a]	2.6± 0.9 (2.5)[a]	(pmol/ml)
Prolactin	4.8± 2.7 (5.0)	7.1± 2.6 (6.2)[a]	4.2± 2.0 (4.3)	(nmol/l)
Cortisol	14.0± 9.0 (11.8)	19.8± 15.1 (12.8)[a]	8.3± 7.1 (5.1)[a]	(nmol/l)
Calcium	2.0± 0.4 (2.1)	2.1± 0.3 (2.3)	2.3± 0.2 (2.3)	(mEq/l)
Magnesium	2.0± 0.3 (2.2)	2.0± 0.4 (2.2)	2.2± 0.1 (2.3)	(mEq/l)
Zinc	2.4± 0.4 (2.4)	2.8± 0.9 (2.9)	2.6± 0.4 (2.6)	(mEq/l)
Oxytocin	7.5± 4.9 (6.7)	13.4± 6.1 (13.0)[a]	10.0± 4.1 (9.3)	(pg/ml)

HVA, homovanillic acid; MHPG, 3-methoxy-4-hydroxyphenylglycol; HIAA, hydroxyindole-acetic acid; AMP, adenosine 5'-monophosphate; GMP, guanosine 3',5'-monophosphate

[a] Significant differences from controls

hulls were introduced: a convex hull of a two-dimensional cloud of points is the region defined by its extreme points and the straight lines between them. To obtain the $(1-p)$ convex hull, a proportion p of the most extreme subjects is eliminated, such that the area covered becomes minimized ($p = 1/16$ in the present study).

Figure 1 provides the two-dimensional representation of all 44 subjects for 17 CSF biochemical parameters obtained by MDS. The normative region, defined by the 15/16 convex hull of the control group, is quite compact and does not contain a single schizophrenic subject. There is one subject of the control group lying amid schizophrenic subjects. With one exception, all untreated schizophrenic patients are close together, separated from the normative region with respect to dimension 2. The group of schizophrenics on neuroleptics, on the other hand, is on the average more distant from the normative region and shows a very wide scatter, indicating a substantial heterogeneity with regard to the CSF parameters.

Our results indicate that a biological heterogeneity between schizophrenic and nonschizophrenic subjects can be detected by the simultaneous analysis of the CSF concentrations of substances related directly or indirectly to the neuronal activity in the brain. The two-dimensional reduction of 17 CSF parameters via MDS followed by the introduction of $(1-p)$ convex hulls (Fig. 1) correctly separated 15 out of 16 controls from the schizophrenic subjects.

It is worth noting that whereas both controls and drug-free schizophrenics were separated in two distinct and relatively compact regions, patients on neuroleptics were heterogeneously distributed in a very wide space, with the largest distances from the normative region. This graphic representation is in line with the data in Table 1: the concentrations of at least seven parameters that were similar in controls and drug-free patients deviated significantly in patients on drugs, suggesting an effect of neuroleptics upon the systems regulating the synthesis and secretion of these substances into the CSF.

In conclusion, our findings suggest that the multidimensional approach used here could provide an important strategy for further studies on schizophrenia with the aim of elucidating the biology of the disease.

Acknowledgments. This work was made possible through the skillful collaboration of Dr. P. Waldmeier (Ciba Geigy, Basel), Prof. Dr. P. Riederer (Ludwig Bolzmann Institute, Vienna), Prof. Dr. H. Cramer (University of Freiburg), Dr. G. P. Reynolds (MRC Neurochemical Pharmacology Unit, Cambridge), Dr. R. E. Lang (University of Heidelberg), and Prof. Dr. J. S. Kim (University of Ulm) in the biochemical determinations. The data analysis was performed with the support of the Sonderforschungsbereich 123 at the University of Heidelberg.

References

Andén NE, Bedard P (1971) Influences of cholinergic mechanisms on the function and turnover of brain dopamine. J Pharm Pharmacol 23:460–462

Asberg M, Träskman L, Thorén P (1976) 5-HIAA in the cerebrospinal fluid. Arch Gen Psychiat 33:1193–1197

Ashcroft G, Crawford T, Ecclestone D, Sharman D, MacDougall E, Stanton J, Binns J (1966) 5-Hydroxyindole compounds in the cerebrospinal fluid of patients with psychiatric or neurological diseases. Lancet ii: 1049–1052

Banki CM, Molnár G (1980) Cerebrospinal fluid 5-hydroxyindoleacetic acid as an index of central serotonergic processes. Psychiat Res 5:23–32

Barbeau A (1962) Pathogenesis of Parkinson's disease: a new hypothesis. Can Med Assoc J 87:802–807

Bartholini G, Stadler H, Lloyd KG (1973) Cholinergic-dopaminergic interactions in the extrapyramidal system. Advanc Neurol 3:233–241

Bartholini G, Stadler H, Gadea-Ciria M, Lloyd KG (1976) The use of the push-pull cannula to estimate the dynamics of acetylcholine and catecholamines within various brain areas. Neuropharmacology 15:515–519

Beckmann H, Reynolds GP, Sandler M, Waldmeier P, Lauber J, Riederer P, Gattaz WF (1982) Phenylethylamine and phenylacetic acid in CSF of schizophrenics and healthy controls. Arch Psychiatr Nervenkr 232:463–471

Beckmann H, Lang RE, Gattaz WF (to be published) Vasopressin-oxytocin in cerebrospinal fluid of schizophrenic patients and normal controls. Psychoneuroendocrinology

Bird ED, Barnes J, Iversen LL, Spokes EG, Mackay AVP, Shepherd M (1977) Increased brain dopamine and reduced glutamic acid decarboxylase and choline acetyltransferase activity in schizophrenia and related psychoses. Lancet ii:1157–1159

Bird ED, Spokes EG, Iversen LL (1979) Brain norepinephrine and dopamine in schizophrenia. Science 204:93–94

Bowers MB (1973) 5-Hydroxyindoleacetic acid (5-HIAA) and homovanillic acid (HVA) following probenecid in acute psychotic patients treated with phenothiazines. Psychopharmacology 28:309–318

Bowers MB, Henninger GR, Gerbode FA (1969) Cerebrospinal fluid 5-hydroxyindoleacetic acid in psychiatric patients. Int J Neuropharmacol 8:255–262

Carroll BJ (1976) Limbic system-adrenal cortex regulation in depression and schizophrenia. Psychosom Med 38:106–121

Crow TJ, Owen F, Cross AJ, Johnstone EC, Joseph MH, Longden A (1980) The dopamine receptor as the site of the primary disturbance in the type I syndrome of schizophrenia. In: Usdin E et al. (eds) Enzymes and neurotransmitters in mental disease. Wiley, New York, pp 559–572

Ebstein RP, Biederman J, Rimon R, Zohar J, Belmaker RH (1976) Cyclic GMP in the CSF of patients with schizophrenia before and after neuroleptic treatment. Psychopharmacology 51:71–74

Farley IF, Price KS, McCullough E, Deck JHN, Hordynski W, Hornykiewicz O (1978) Norepinephrine in chronic paranoid schizophrenia: above-normal levels in limbic forebrain. Science 200:456–458

Gasser T, Möcks J (1983) Graphical representation of multidimensional EEG data and classificatory aspects. Electroencephalogr Clin Neurophysiol 55:609–612

Gattaz WF, Gattaz D, Beckmann H (1982a) Glutamate in schizophrenics and healthy controls. Arch Psychiat Nervenkr 231:221–225

Gattaz WF, Waldmeier P, Beckmann H (1982b) CSF monoamine metabolites in schizophrenic patients. Acta Psychiat Scand 66:350–360

Gattaz WF, Cramer H, Beckmann H (1983a) Low CSF concentration of cyclic GMP in schizophrenia. Br J Psychiatry 142:288–291

Gattaz WF, Riederer P, Reynolds GP, Gattaz D, Beckmann H (1983b) Dopamine and noradrenalin in the cerebrospinal fluid of schizophrenic patients. Psychiat Res 8:243–250

Gattaz WF, Kattermann R, Gattaz D, Beckmann H (1983c) Magnesium and calcium in the CSF of schizophrenic patients and healthy controls: correlations with cyclic GMP. Biol Psychiat 18:935–939

Gomes UCR, Shanley BC, Potgieter L, Roux JT (1980) Noradrenergic overactivity in chronic schizophrenia: evidence based on cerebrospinal fluid noradrenaline and cyclic nucleotide concentrations. Br J Psychiatry 137:346–351

Heinrich K (1976) Psychopharmaka in Klinik und Praxis. Thieme, Stuttgart

Helmchen H, Hippius H (1967) Depressive Syndrome im Verlauf neuroleptischer Therapie. Nervenarzt 38:455–458

Johnson JL (1972) Glutamic acid as a synaptic transmitter in the nervous system. A review. Brain Res 37:1–19

Kizer JS, Youngblood WW (1978) Neurotransmitter systems and central neuroendocrine regulation. In: Lipton MA, DiMascio A, Killam KF (eds) Psychopharmacology: a generation of progress. Raven, New York, pp 465–486

Lake CR, Sternberg DE, Van Kammen DP, Ballenger JC, Ziegler MG, Post RM, Kopin IJ, Bunny WE (1980) Schizophrenia: elevated cerebrospinal fluid norepinephrine. Science 207:331–333

Langer DH, Brown GL, Docherty JP (1981) Dopamine receptor supersensitivity and schizophrenia: a review. Schizophrenia Bull 7:208–223

McGlashan TH, Carpenter WT (1976) Postpsychotic depression in schizophrenia. Arch Gen Psychiat 33:231–239

Murphy BEP, Cosgrove JB, McIlguham MC, Pattee CJ (1967) Adrenal corticoid levels in human cerebrospinal fluid. Can Med Assoc J 97:13–17

Persson T, Roos B-E (1969) Acid metabolites from monoamines in cerebrospinal fluid of chronic schizophrenics. Br J Psychiatry 115:95–98

Post RM, Kotin J, Goodwin FK, Gordon EK (1973) Psychomotor activity and cerebrospinal fluid amine metabolites in affective illness. Am J Psychiatry 130:67–72

Rowlands GJ, Roberts PJ (1980) Activation of dopamine receptors inhibit calcium-dependent glutamate release from corticostriatal terminals in vitro. Eur J Pharmacol 62:241–242

Sedvall G, Alfredsson G, Bjerkenstedt L, Eneroth P, Fyrö B, Härnryd C, Swahn CG, Wisel A, Wode-Helgodt B (1975) Selective effects of psychoactive drugs on levels of monoamine metabolites and prolactin in cerebrospinal fluid of psychiatric patients. In: Airaksinen M (ed) Proceedings of the sixth congress of pharmacology. Forssan Kirjapaino y oy, Forssa, vol 3, pp 255–267

Sedvall GC, Wode-Helgodt B (1980) Aberrant monoamine metabolite levels in CSF and family history of schizophrenia. Arch Gen Psychiat 37:1113–1116

Smith CC, Tallman JF, Post RM (1976) An examination of baseline and drug induced levels of cyclic nucleotides in the cerebrospinal fluid of controls and psychiatric patients. Life Sci 19:131–136

Snyder S (1976) The dopamine hypothesis of schizophrenia: focus on the dopamine receptor. Am J Psychiatry 133:197–202

Snyder SH, Banergee SP, Yamamma HI, Greenberg D (1974) Drugs, neurotransmitters and schizophrenia. Science 184:1243–1253

Spitzer RL, Endicott J, Robins E (1975) Research diagnostic criteria. Instrument No. 58. New York State Psychiatric Institute, New York

Stadler H, Lloyd KG, Gadea-Ciria M, Bartholini G (1973) Enhanced striatal acetylcholine release by chlorpromazine and its reversal by apomorphine. Brain Res 55:476–480

Stadler H, Gadea-Ciria M, Bartholini G (1975) In vivo release of endogenous neurotransmitters in cat limbic regions: effect of chlorpromazine and of electrical stimulation. Naunyn-Schmiedebergs Arch Pharmacol 288:1–6

Stein L, Wise CD (1971) Possible etiology of schizophrenia: progressive damage to the noradrenergic reward system by 6-hydroxydopamine. Science 171:1032–1036

Sternberg DE, Van Kammen DP, Lake CR, Ballenger JC, Marder SR, Bunney WE (1981) The effect of pimozide on CSF norepinephrine in schizophrenia. Am J Psychiatry 138:1045–1051

Trabucchi M, Cheney DL, Racagne G, Costa E (1975) In vivo inhibition of striatal acetylcholine turnover by L-dopa, apomorphine and (+)-amphetamine. Brain Res 85:130–134

Winblad B, Bucht G, Gottfries CG, Roos B-E (1979) Monoamines and monoamine metabolites in brain from demented schizophrenics. Acta Psychiat Scand 60:17–28

Abbreviations

A	adrenaline	5-HT	5-hydroxytryptamin (serotonin)
ACh	acetylcholine	5-HTP	5-hydroxytryptophan
ACTH	adrenocorticotropic hormone	HVA	homovanillic acid
³H-ADTN	(1)-6,7-dihydroxy-2-aminotetralin	ICD	international classification of diseases
AIP	acute intermittent porphyria		
ALA	aminolevulinic acid	LH	luteinizing hormone
AMDP	Arbeitsgemeinschaft für Methodik und Dokumentation in der Psychiatrie	LP	lumbar puncture
		LSD	lysergic acid diethylamide
		MAO	monoamine oxidase
ATP	adenosine triphosphate	MHPG	3-methoxy-4-hydroxyphenyl-glycol
BPRS	Brief Psychiatric Rating Scale		
CAT	choline acetyltransferase	MMPI	Minnesota Multiphasic Personality Inventory
CBF	cerebral blood flow		
CCK	cholecystokinin	NA	noradrenaline
CGI	Clinical Global Impression Scale	NE	norepinephrine
CMV	cytomegalovirus	NF	norfenfluramine
CNS	central nervous system	NLs	neuroleptic drugs
CPRS	Comprehensive Psychiatric Rating Scale	NT	neurotensin
		PEG	pneumencephalogram
CSF	cerebrospinal fluid	PET	positron emission tomography
CT	computertogram	PRL	prolactin
DA	dopamine	PST	phenolsulphotransferase
DBH	dopamine-β-hydrolase	RDC	Research Diagnostic Criteria
DOPAC	dihydroxyphenylacetic acid	SAH	S-adenosylhomocysteine
DSM-III	Diagnostic and Statistical Manual of Mental Disorders	SAM	S-adenosylmethionine
		SHMT	serine hydroxymethyl-transferase
DST	dexamethasone suppression test		
EEG	echoencephalogram	SRIF	somatostatin
FRC	Feighner's Research Criteria	THF	tetrahydrofolic acid
FSH	follicle-stimulating hormone	TIQs	tetrahydroisoquinolines
GABA	gamma-aminobutyric acid	TRH	tyrotropin-releasing hormone
GAD	glutamate decarboxylase	TRY	tryptophan
GH	growth hormone	TSH	tyroid-stimulating hormone
GMP	cyclic guanosine 3'5'-mono-phosphate	VBRs	ventricle–brain ratios
		VEPs	visually evoked cortical potentials
5-HIAA	5-hydroxyindoleacetic acid		
HPLC	high-pressure liquid chroma-tography	VIP	vasoactive intestinal poly-peptide

Subject Index